Computer Mathematics Concepts

Elizabeth Chu   James Fulton Theodore Koukounas

Suffolk County Community College

Copyright Page:

All rights reserved. No part of this book may be reproduced, in any form or by any means, without permission in writing from the publisher.

Table of Contents

Chapter 1 The Basics of Computer Programming, Numbers, and Operations *1*

Section 1.1: Computer Programming with BASIC *2*

Section 1.2: Real Numbers and Order *10*
    Objective A: The Number System *10*
    Objective B: Entering Numerical Data into the Computer *14*
    Objective C: Order and Real Numbers *16*

Section 1.3: Operations on Real Numbers *19*
    Objective A: Absolute Value *19*
    Objective B: Addition and Subtraction of Real Numbers *21*
    Objective C: Inverse Operations and Inverse Operators *24*
    Objective D: Multiplication and Division of Real Numbers *25*
    Objective E: Exponents of Real Numbers *28*

Section 1.4: Additional Operations and Properties of Real Numbers *37*
    Objective A: Roots and Radicals of Real Numbers *37*
    Objective B: Properties of Real Numbers *39*
    Objective C: Percents *43*
    Objective D: Order of Operations *46*

Section 1.5: Decimal Representation of Real Numbers *55*
    Objective A: Fixed-Point Numbers *55*
    Objective B: Floating-Point and Scientific Notation *57*

Section 1.6: Two Additional Common Operators: MOD and INT *60*

Chapter 2   Numbers and Computers *63*

Section 2.1: Computer Terminology and Units of Measurements *74*
    Objective A: Bits and Bytes *68*
    Objective B: Measurements of Time *81*
    Objective C: Processor Speed *82*
    Objective D: File Transfer Rate *83*
    Objective E: Disk Storage Capacity *84*
    Objective F: Mean Time Between Failure *85*

Section 2.2: Working With Numbers in a Computer *89*
        Objective A: Representing Real Numbers in a Computer *89*
        Objective B: Approximation and Error *91*
        Objective C: Absolute and Relative Error (Optional) *94*
        Objective D: Conversion, Accumulation and Subtraction Error (Optional) *96*

Section 2.3: Decimal, Binary, Octal, and Hexadecimal Numbers *103*
        Objective A: Converting Binary, Octal, and Hexadecimal to Decimal *104*
        Objective B: Converting Decimal to Binary and Hexadecimal *112*
        Objective C: Converting Between Binary, Octal and Hexadecimal *119*

Section 2.4: Fractional Numbers *124*
        Objective A: Converting Fractional Binary and Hexadecimal to Decimal *125*

Section 2.5: Basic Arithmetic in Binary and Hexadecimal Numbers *129*
        Objective A: Addition of Unsigned Integers *129*
        Objective B: Subtraction of Unsigned Integers *132*

Section 2.6: Computer Arithmetic *136*
        Objective A: Fixed-Bit Representation of Signed Integers and the Two's Complement Method *136*
        Objective B: Addition and Subtraction of Signed Integers *141*
        Objective C: Why the Two's Complement Method Works (Optional) *144*

Section 2.7: Applications of Binary and Hexadecimal *147*
        Objective A: ASCII and Unicode *147*
        Objective B: Memory Identification *148*

Chapter 3 Logic and Computers *150*

Section 3.1: Logical Statements and Connectives *151*
        Objective A: Simple Statements *151*
        Objective B: Connectives and Compound Statements *152*

Section 3.2: Conjunction and Disjunction *157*
        Objective A: Disjunction and the Logical Operator OR *157*
        Objective B: Conjunction and the Logical Operator AND *159*
        Objective C: Programming Applications *162*

Section 3.3: Negation of Statements *166*
        Objective A: Negation and the Logical Operator NOT *166*
        Objective B: Quantifiers and Negation of Quantified Statements *167*

Section 3.4: Tree Diagrams and Truth Tables *172*

Section 3.5: Conditional and Biconditional *180*
        Objective A: The Conditional Statement *180*
        Objective B: Programming Applications and the Conditional *183*
        Objective C: Biconditional of Statements (Optional) *188*
        Objective D: Overview of Truth Tables *190*

Section 3.6: Logical Equivalences *194*
        Objective A: Logically Equivalent Statements *194*
        Objective B: De Morgan's Laws *195*
        Objective C: Contrapositive, Converse and Inverse of a Conditional Statement *196*
        Objective D: Distributive Properties(Optional) *198*

Section 3.7: Applications with Electric Circuits *201*

Chapter 4   Relations, Functions, and Subroutines *205*

Section 4.1: Relations and Functions *206*
        Objective A: Defining Relations *206*
        Objective B: Defining Functions *209*
        Objective C: Functional Notation *211*

Section 4.2: Some Special Mathematical Functions *218*
        Objective A: Square Root Function *218*
        Objective B: Absolute Value Function *219*
        Objective C: MOD Function *219*
        Objective D: Exponential Functions *220*
        Objective E: Properties of Exponential Functions (Optional) *224*
        Objective F: Logarithmic Functions *225*
        Objective G: Properties of Logarithmic Functions (Optional) *234*

Section 4.3: Additional Built-in BASIC Functions (Optional) *239*

Section 4.4: Subroutines and User Defined Functions *242*
        Objective A: Subroutines *242*
        Objective B: User Defined Functions *244*

Section 4.5: Repetition Structures *249*
        Objective A: FOR-NEXT Loop *249*
        Objective B: DO-WHILE Loop *251*
        Objective C: DO-UNTIL Loop *253*

Section 4.6: Recursion and Recursive Functions *259*

Chapter 5 Subscripts, Vectors, Matrices and Arrays *267*

Section 5.1: One-Dimensional Arrays: Vectors *268*
        Objective A: Size of a Vector and One-Dimensional Arrays *268*
        Objective B: Equality of Vectors *271*

Section 5.2: Vector Arithmetic *274*
        Objective A: Addition and Subtraction of Vectors *274*
        Objective B: Multiplication of Vectors *276*
        Objective C: One-Dimensional Arrays and Computers *280*
        Objective D: Geometric Interpretation of Vectors (Optional) *282*

Section 5.3: Two-Dimensional Arrays: Matrices *288*
        Objective A: Order of a Matrix and Two-Dimensional Arrays *288*
        Objective B: Equality of Matrices *291*

Section 5.4: Matrix Arithmetic *293*
        Objective A: Addition and Subtraction of Matrices *293*
        Objective B: Multiplication of Matrices *293*
        Objective C: Identity Matrix *299*
        Objective D: Two-Dimensional Arrays and Computers *300*
        Objective E: An Interpretation of Matrix Multiplication (Optional) *303*
        Objective F: Inverse of a Matrix (Optional) *305*

Section 5.5: Summation and Product Operators *313*
        Objective A: Summation Operator *313*
        Objective B: Product Operator *318*

Chapter 6 Set Theory – Preliminary Concepts *322*

Section 6.1: Sets and Set Notation *323*
        Objective A: Describing Sets *323*
        Objective B: The Universal and Empty Set *327*

Section 6.2: Subsets and Proper Subsets *330*
        Objective A: Subsets *330*
        Objective B: Proper Subsets *332*
        Objective C: Complement Operation of Sets *334*

Section 6.3: Comparison and Cardinality of Sets *339*
        Objective A: Set Equality *339*
        Objective B: Cardinality of Sets *340*
        Objective C: Set Equivalence *341*
        Objective D: Finite and Infinite Sets *342*

Section 6.4: Set Operations *346*
        Objective A: Union of Sets *346*
        Objective B: Intersection of Sets *348*
        Objective C: Difference of Sets *353*
        Objective D: Applications of Set Cardinality *354*

Appendix A: BASIC Statements *361*

Appendix B: Geometric Formulas *372*

Appendix C: ASCII Table *376*

Appendix D: Answers to Select Exercises *380*

Preface

In 1996, the Computer Science Department began a new curriculum called Computer Technology which is the current Information Technology program. The Computer Science Department requested the Mathematics Department to design and develop a mathematics course specifically for this curriculum.

Josephine Freedman of the Computer Science Department and Elizabeth Chu of the Mathematics Department collaborated and developed a new mathematics course called Computer Mathematics Concepts. The course is taught in a classroom where computers are used to demonstrate mathematical concepts.

For many years, an appropriate textbook covering all the topics needing to be covered could not be found. As a result, we decided to collaborate to create this text. We wanted a textbook that incorporates mathematical concepts and some simple BASIC programming.

Acknowledgements

We would like to thank the following colleagues who have contributed to this project since its inception: Beverly Broomell, Leslie Buck, Ed Rodriguez, Laura Smith, and Masako Stampf for their contributions to the lab manual that preceded this text.

We would also like to thank Marilyn Campo, for her hard work and dedication; her meticulous proofreading served to assure the accuracy of the textbook. In addition, many thanks to Adrienne Chu for working on the solutions to the exercises as well as the photo for the book cover.

We would like to thank the following reviewers for their time and suggestions: Bruce Barton, Xingbin(Ben) Chen, Maureen O'Grady, and Debra Wakefield.

We also wish to thank Michael Jones for his valuable advice, technical assistance, and computer expertise that helped to make this book possible.

Finally, we would especially like to thank Dennis Reissig, who spent many hours reviewing the text and gave us indispensable guidance as well as invaluable pedagogical insights.

Elizabeth Chu   *James Fulton*   *Theodore Koukounas*

June, 2004

Chapter 1

The Basics of Computer Programming, Numbers, and Operations

A computer is essentially a fast and powerful calculating machine. It takes an input, processes it and produces some output. For example, a computer might take a list of employees working for a company, along with their hourly pay and the number of hours they each worked that week, and compute their weekly salaries. It might also respond to the commands and inputs of a young "gamer" as he or she tries to conquer some virtual environment. It could also take temperatures, pressures and wind patterns and make predictions about weather conditions for meteorologists. It might read a CD or a DVD, process the information, and output a song or a movie. The possibilities are endless!

Although these examples seem so different, there are some fundamental, underlying connections that they all have. They all involve the input, processing and output of information. Computers process information in the form of symbols, just as humans do. These symbols might represent letters, words, actions or numbers. They are the fundamental building blocks of a computer program. Thus, a thorough understanding of each is critical.

Before we can understand and fully utilize the power of the computer, we need to know how to work with numbers. In this chapter, we focus our attention on numbers and operations involving numbers that everyone working with computers needs to be familiar with. Numbers are the most fundamental and important part of the computer. We must have a good "number-sense."

We will introduce some elementary concepts in computer programming, especially as it relates to the mathematics we present in this chapter.

After completing this chapter you should have obtained a good "number sense." More specifically, you should be able to:
- distinguish between the various sets of numbers; natural, whole, integer, rational, irrational and real,
- convert to and work with decimal and floating point numbers,
- order a set of numbers,
- perform various operations on real numbers, such as: addition, subtraction, multiplication, division, exponentiation, finding the absolute value, as well as any combination of the above,
- understand and apply the order of operations,

- understand roots and radical notation,
- understand how numbers are stored on a computer (fixed-point and floating-point),
- work with the MOD, ABS, INT, SQR operators, and
- comfortably implement all of the above mathematics in a simple **BASIC** program.

Section 1.1: Computer Programming with BASIC

Before taking a look at numbers and their properties, we will need to understand the context in which they will be used. For us, that is the computer program. A **computer program** consists of a set of instructions that tells the computer how to perform a specific task. The program is usually written to solve a problem.

A program typically has three main features: **input** of data, **process** of data, and **output** of data.

1. **Input** - - data entered through a computer keyboard, floppy disk, etc.
2. **Process** - - could be calculation of data, sort data, or perform search, etc.
3. **Output** - - either printed out on paper or displayed on the monitor.

Flowchart - - graphical display of the steps in a program

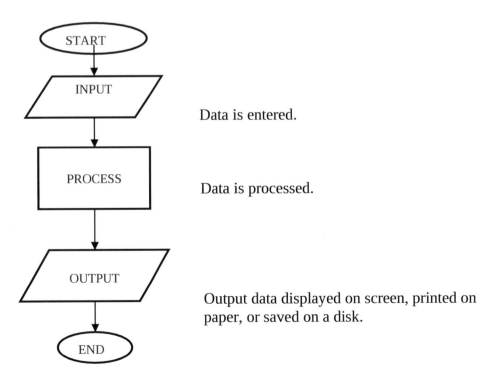

There are many programming languages. Some examples are C, C++, Java and BASIC. These programming languages are considered high-level (or symbolic) languages because the codes are written such that they are easy for programmers to understand.

A version of **BASIC** (Beginner's All-purpose Symbolic Instruction Code) will be used in this course. **QBASIC** was used in the past, but any version of **BASIC** can be used. The programs that you see in this text should run in most versions of **BASIC** with possibly some minor modifications. This programming language is chosen because it is easy to learn and use.

Correct **syntax** and **logic** are essential for a program to run successfully. **Syntax** (coding rule) is used to form valid instructions in writing a program. A missing symbol, such as a quotation mark, could result in the program not running at all. **Logic** is also an important part of writing a program. You may have correct syntax all through your program, however, if the logic of the sequence of your statements is incorrect, there will be inconsistent results and/or other problems when you run your program.

In this section, we will learn some fundamental programming techniques used to perform arithmetic calculations. Simple **BASIC** statements will be introduced. We shall see the output of each of these simple programs. We shall also learn how to trace a program to see what the output is going to be.

Assignment Statements and Variables

The first example below simply adds two numbers much like a calculator. But the calculator is "pre-programmed" while the program below tells the computer to add these two numbers.

Example 1:

```
Line#  code
10     CLS
20     A = 3
30     B = 4
40     C = A + B
50     PRINT A,B,C
60     END
```

The program above contains six lines of code. Line numbers are placed at the beginning of each statement so that it is easy for us to refer to any specific line. This is not a requirement for all **BASIC** compilers or interpreters. Some versions of **BASIC** may not accept line numbers. We use them in a couple of examples for the purpose of discussion only. The remaining programs in this text are written without the line numbers.

The **CLS** statement on line 10 simply means clear the screen before executing the next statement. The **END** statement on line 60 means it is the end of the program.

Lines 20, 30, and 40 are all called **assignment statements**. Line 20 assigns the number 3 to the variable A, line 30 assigns the number 4 to the variable B. Line 40 assigns the sum of the contents of A and B to variable C. The content of C is 7 after line 40 is executed. Variables will be discussed in more detail in section 1.7.

Line 50 indicates that the contents of A, B, and C are to be printed. The commas separating the variables tell the computer to display the content of each variable in preset print fields. These print fields are usually 14 spaces wide. To conserve space in our textbook, we will always use a print field of 7 spaces. If variables A, B, and C were separated with semi-colons, then the contents will be displayed very close to each other with only one space between the numbers.

Hence, when we run this program, we should see an output of :

       OUTPUT:
  3   4   7

We can also enter letters of the alphabet, digits or symbols such as semicolon and period into the computer. These are called **character strings** or simply **strings**. The variable name used to name a character string is followed by a $. The string must be enclosed in quotation marks. The following is an example of how a **character string assignment** can be done.

***Example 2*:**

```
Line#code
10   CLS
20   MESSAGE$ = "Hello. This is an interesting course."
30   PRINT MESSAGE$
40   END
```

Notice there is a **$** after the word MESSAGE in line 20. This means that what is being assigned to the variable MESSAGE consists of characters or string. This program prints a message, so the output of this program is just the following:

OUTPUT:
Hello. This is an interesting course.

We have seen two simple programs that showed how the assignment statements are used. More about this topic will be discussed in section 1.7. Now we'll introduce a new BASIC statement that is very useful in writing programs, especially long programs. It is the **REM** statement.

REM stands for "remark" or comment. It is used when you want to document or make comments in your program for yourself or other people reading the program. This line will not be executed. The single quotation mark (') symbol can be used instead of **REM.**

For example,
```
      REM   Add the numbers A and B
```
indicates that the two numbers in A and B are to be added. Or for the line below
```
      A = Base*Height      ' calculate the area
```

The remark on the right explains what the statement to its left does. Here, it is calculating the area. Now, let us see an example:

Example 3: Enter the program in your computer and RUN the program to see what the output will be.

```
REM   print the result after multiplying the values stored
      in num1 and num2
num1 = 4
num2 = 2
PRINT   num1*num2
END
```

In this program the number 4 is assigned to the variable num1 and the number 2 is assigned to num2. Then the product of the contents of num1 and num2 is printed. The result will be displayed on the screen.

OUTPUT:
8

Well documented programs are very useful since it makes it easier for not only the programmer but other people who might have to work with the program.

Tracing a Program and Displaying the Output

One very useful tool that is used in programming is called tracing. We look at each line of the program code and determine what the value of each variable is at that time, what operations are being performed, and what the output will look like.

Example:

We have completed the following table by filling in the value of each variable after each line of the program is executed. Then placed the output that is displayed on the screen in the box below.

	Program	NUM1	NUM2	NUM3
10	CLS			
20	NUM1=4	4		
30	NUM2=6	4	6	
40	NUM3= NUM1+NUM2	4	6	10
50	PRINT NUM1,NUM2,NUM3	4	6	10
60	NUM1=NUM2*NUM3	60	6	10
70	PRINT NUM1-NUM2	60	6	10
80	END			

(variables: NUM1, NUM2, NUM3)

- Lines 20 and 30 assign the numbers 4 and 6 to NUM1 and NUM2 respectively. So you see the two numbers placed in the boxes provided to the right of the code. Notice that line 30 only changes NUM2; NUM1 remains the same as it was in line 20.

- Line 40 assigns the sum of NUM1 and NUM2 to NUM3. So the NUM3 should have 10 in it. NUM1 and NUM2 are unchanged.

- Line 50 states to print NUM1, NUM2 and NUM3. This means that whatever is in each variable will be printed, in this case to the screen or monitor. Notice the first line in the output box below has the three numbers 4, 6, and 10.

- Line 60 assigns the product of the numbers in NUM2 and NUM3 to NUM1. So NUM1 now has 6 times 10 which is 60.

- Line 70 prints the difference between NUM1 and NUM2. The contents of NUM1 and NUM2 are still the same as line 60.

- Line 80 terminates the program.

   OUTPUT:
   4   6   10
   54

COMPUTER ACTIVITY

Student Name_____ Section_____ Date_____

1. Complete the table by filling in the value of each variable after each line of the program is executed. Then place the output that is displayed on the screen in the box below.

Program	NUM1	NUM2	NUM3
CLS			
NUM1=2			
NUM2=5			
NUM3= NUM1- NUM2			
PRINT NUM1,NUM2,NUM3			
NUM1=2*NUM2*NUM3			
PRINT NUM1+NUM2			
NUM2=NUM2+NUM3			
PRINT NUM1,NUM2,NUM3			
END			

What should your screen look like when you RUN the program?

OUTPUT:

COMPUTER ACTIVITY

Student Name_____ Section_____ Date_____

2. Complete the table by filling in the value of each variable after each line of the program is executed. Fill the OUTPUT box with what you will see in your screen.

Program	balance	deposit	withdraw	rate
balance = 1000	1000			
deposit = 200	1000	200		
withdraw = 525	1000	200	525	
rate = 0.05	1000	200	525	0.05
balance = balance + rate*balance	1050	200	525	0.05
balance = balance + deposit	1250	200	525	0.05
balance = balance - withdraw	725	200	525	0.05
PRINT balance	725	200	525	0.05
PRINT balance + deposit	725	200	525	0.05

What should your screen look like when you RUN the program?

OUTPUT:

```
725
925
```

Section 1.2: Real Numbers and Order

Now that we know how to run a simple program, let's focus on how we input information into a computer. One of the most basic ways to input or output data into or out of a computer is with numbers. As you probably recall from early lessons in mathematics, there are many different "types" of numbers. You should be able to distinguish between these "types" when using or working on a computer. There are essentially six groupings of numbers that will be important to us: Natural, Whole, Integer, Rational, Irrational and Real. How they are related to each other is the subject of this next subsection.

Objective A: The Number System

The number system we use today was developed over many centuries. As with all developments, its introduction was largely based upon its necessity to solve problems. Counting or natural numbers were the first numbers to be developed. The natural numbers are represented by the bold capital letter **N**, and can be written as a list of numbers such as 1, 2, 3, 4, 5 and 6. Obviously the list goes on and on without end. We represent this list with the following mathematical notation:

$$\mathbf{N} = \{1, 2, 3, 4, 5, 6, \ldots\}$$

The braces "{ }" represent the fact that we are talking about a "set" of numbers. A set will be given a more complete definition later on in this book, but for now we simply take a set to mean a collection of numbers. The "…" is called an ellipsis and means that the list goes on and does not have an end.

Adding zero to the set of natural numbers gives the set of whole numbers, **W**.

$$\mathbf{W} = \{0, 1, 2, 3, 4, 5, \ldots\}$$

The negative of the natural numbers, along with zero and the set of natural numbers, make up the set of integers, **Z**, denoted for zahl which is German for integers. We cannot use **I** for integers because it is reserved for the Irrational numbers.

$$\mathbf{Z} = \{\ldots, -5, -4, -3, -2, -1, 0, 1, 2, 3, 4, 5, \ldots\}$$

A very useful aid in understanding our number system is through the use of a number line. A number line is simply a horizontal line that continues endlessly in both directions as represented by the arrows in the figure below.

We have placed the integers onto the number line a unit distance apart. By convention we put the "larger" numbers to the right of the "smaller" numbers.

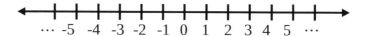

In between the integers on the number line there are gaps. We can begin to fill in these gaps by introducing a set of numbers called Rational Numbers.

Rational Numbers, Q, are defined as the set of all numbers that can be formed as a quotient (ratio, fraction) of two integers, where the denominator can never be zero. The set of rational numbers contains the set of integers since the denominator can be 1.

Examples of rational numbers expressed as the quotient of two integers::

$$\frac{3}{4}, -\frac{1}{3}, \frac{4}{7}, \frac{15}{3}=5, \frac{-21}{2}=-10\frac{1}{2}, 7 \text{ since 7 can be written as } \frac{7}{1}$$

Rational numbers can also be represented as either terminating or repeating decimals.

Examples of rational numbers that are terminating decimals:

$$\frac{3}{4}=0.75, \frac{3}{5}=0.6, \frac{15}{3}=5.0$$

Examples of rational numbers that are repeating decimals:

$$-\frac{1}{3}=-0.33333...=-0.\overline{3}, \frac{17}{99}=171717...=0.\overline{17}, \frac{4}{7}=0.571428571428...=0.\overline{571428}$$

Notice that we can place a bar $^-$ over the digits that are being repeated in a repeating decimal.

Surprisingly, the set of Rational Numbers does not fill up all of the spaces on the number line. It turns out that there are gaps between rational numbers and these gaps are filled by a new set of numbers that are not rational. They are the Irrational Numbers, **I**. The prefix "ir" in irrational is from the Latin meaning "not." Thus, it means "not-rational," i.e. not a ratio of integers.

If we consider the decimal representation of a number, it turns out that an irrational number cannot be represented using either a terminating or a repeating decimal number.

Examples of irrational numbers and their decimal equivalents:

$\sqrt{2} = 1.41421356237...$, $\pi = 3.1415926535...$, $-\sqrt{13} = -3.6055512754...$,

$e^2 = 7.389056098...$, $\sqrt[4]{3} = 1.3160740129...$

All these numbers are irrational since a terminating or repeating decimal pattern is not possible. Thus, the exact representation of irrational numbers can only come from their symbolic form, i.e. $\sqrt{2}$ instead of 1.4142. We will see that because of this property, we can never represent an irrational number on a computer. We can only use a rational approximation.

The set of irrational numbers combined with the set of rational numbers does fill in every space on the number line, and this combined set of numbers is called the set of real numbers.

The relationship between the various sets of numbers is illustrated in the figure below. We see that the natural numbers are also in the set of the whole numbers, which are also in the set of the integers, which are also in the set of the rational numbers.

The irrational numbers are a set that has no commonality with the rational numbers. The rational numbers together with the irrational numbers make up the set of real numbers, **R**.

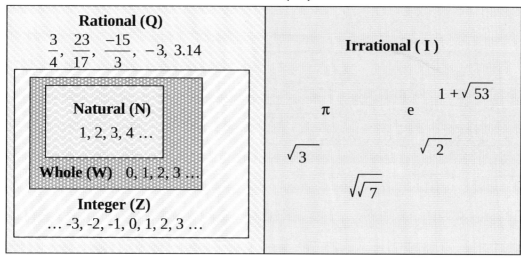

Example 1: Given the following set of numbers below, list the numbers that are
a) Natural numbers
b) Whole numbers
c) Integers
d) Rational numbers
e) Irrational numbers

$$\left\{ 3.14, -5, 7.33\overline{3}, -0.1, \frac{7}{3}, -0.632, \sqrt{13}, 4.96, 3, \pi^2 \right\}$$

Solution: a) Natural numbers: $\{3\}$

b) Whole numbers: $\{3\}$

c) Integers: $\{-5, 3\}$

d) Rational numbers: $\left\{ 3.14, -5, 7.33\overline{3}, -0.1, \frac{7}{3}, -0.632, 4.96, 3 \right\}$

e) Irrational numbers: $\left\{ \sqrt{13}, \pi^2 \right\}$

Example 2: Given the following set of numbers below, list the numbers that are
a) Natural numbers
b) Whole numbers

c) Integers
d) Rational numbers
e) Irrational numbers

$$\left\{0, -3, 4.5, \sqrt{5}, -0.2, \sqrt{36}, \frac{22}{7}, 2.120120012...\right\}$$

Solution: a) Natural numbers: $\left\{\sqrt{36} = 6\right\}$

b) Whole numbers: $\left\{0, \sqrt{36}\right\}$

c) Integers: $\left\{0, -3, \sqrt{36}\right\}$

d) Rational numbers: $\left\{0, -3, 4.5, -0.2, \sqrt{36}, \frac{22}{7}\right\}$

e) Irrational numbers: $\left\{\sqrt{5}, 2.120120012...\right\}$

NOTE: 2.120120012… is not a repeating nor terminating decimal. There is no pattern that repeats.

Objective B: Entering Numerical Data into the Computer

Now that we are familiar with the different types of numbers, let's look at how we enter numerical data into the computer. In **BASIC**, there are several ways to enter numbers into the computer. We can simply use the **assignment statement**, use the **INPUT** statement, or use the **READ/DATA** statement. Here, we introduce the **INPUT** statement which is of the form:

INPUT ["prompt"];*variable*
The prompt tells the user to enter data or information needed to continue executing the program. The data entered is assigned to the variable.

The brackets [ ] around the "prompt" are optional, but if an INPUT statement does not have a prompt, then a question mark "?" will appear when that line is executed. This is not very helpful since the user does not know what information is needed. Try to enter the following line in the computer and see what is displayed.

```
INPUT Grade
```

When you run this one line program, you will only see a "?" on the screen and unless you are the programmer, you won't know what information is wanted.

Now, type and run the following simple program. See what the output will be for the data given.

```
CLS
INPUT "Type in the numerator of the rational number."; p
INPUT "Type in the denominator of the rational number."; q
PRINT "The decimal representation of the rational number
      is : "; p/q
END
```

1. Enter 15 for p and 3 for q. The output should be 5.
2. Enter 2 for p and 3 for q. The output should be 0.6666667. (rounded number.)
3. Enter 10 for p and 0 for q. You will see an error message since we cannot divide by 0.

The INPUT statement is one of the easiest ways to get numerical data into a computer. It is possible to enter in two or more data values with one command, but care must be taken to alert the user that more than one items are expected, otherwise the program may cause problems for the user.

Try the following program:

```
CLS
PRINT "Type in the numerator and the denominator of the
       rational number."
INPUT "Data must be entered in the form: number, number";
      p,q
PRINT "The decimal representation of the rational number is
      : "; p/q
END
```

When you run this program, see what happens when you only type one number and hit return. Experiment with this program, so that you are comfortable with how the INPUT command works.

Input the number 22 for p and 7 for q. Make sure to enter both numbers separated with a comma. What irrational number does this rational number approximate?
Change the values of p and q to see the decimal equivalents of the rational numbers you've entered. Notice you always get either a terminating or a repeating decimal pattern.

Assignment statements and **READ/DATA** statements will be discussed in more detail in section 1.7.

Objective C: Order and Real Numbers

In placing the real numbers on a number line, we are able to establish the property of **order** between the various numbers. We placed the number 2 to the right of the number 1, since 2 is greater in value than 1, or we say that 2 comes after 1 on the number line. In a similar way we place a negative three (-3) to the left of -2 on the number line, indicating that –3 is less than –2.

Rather than writing out a long series of words to help us describe this order, we have adopted the following **relational** symbols for real numbers a and b.

Symbol	Meaning
$a < b$	a is less than b
$a > b$	a is greater than b
$a = b$	a is equal to b
$a \neq b$	a not equal to b
$a \leq b$	a is less than or equal to b
$a \geq b$	a is greater than or equal to b

We make mathematical statements by combining the relational symbols with numbers. The truth or falsity of these statements can then be determined. For example consider the following:

Examples : Give the meaning of each of the following statements and find whether it is true or false.

Statement	Meaning	Truth Value
$2 < 6$	two is less than six	True
$5 = 7$	five is equal to seven	False
$-3 > -2$	negative three is greater than negative two	False
$-4 \leq -4$	negative four is less than or equal to negative four	True

The *order* of a pair of numbers is based upon their relative positions on a number line. This is why the statement $-3 > -2$ is considered false. Even though -3 can be considered to be a larger debt than -2, it is, however, to the left of -2 on the number line, therefore it is a "smaller" number (i.e. it is less than -2). This can be seen from the graph below where -3 is to the left of -2.

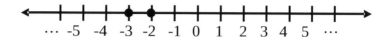

EXERCISES 1.2

For each of the following sets of numbers, list the numbers that are
      a) Natural numbers
      b) Whole numbers
      c) Integers
      d) Rational numbers
      e) Irrational numbers

1. $\left\{3.4, -5, 52, 0, \dfrac{3}{8}, -0.632, \sqrt{7}, 496, -3.25, -\pi\right\}$

2. $\left\{-6, 1.5, \dfrac{1}{2}, \sqrt{25}, -0.62, \sqrt{3}, 92, -8.33\right\}$

3. $\left\{23, -1, 1.2, \dfrac{3}{7}, -\dfrac{5}{6}, \sqrt{11}, 567, 0, \pi, -1.34\right\}$

What is the output of the following program segments?

4.
```
CLS
INPUT "Enter a number."; a
INPUT "Enter a number."; b
PRINT "The sum of a and b is "; a + b
END
Use  a = 6 and b = 9.
```

5.  ```
    CLS
    INPUT "Enter a number."; a
    INPUT "Enter a number."; b
    PRINT "The product of a and b is "; a * b
    END
    ```
    Use a = 5 and b = 7.

6.  ```
    CLS
    INPUT "Enter a number."; a
    INPUT "Enter a number."; b
    PRINT "The difference of a and b is "; a - b
    END
    ```
    Use a = 15 and b = 4.

Write the meaning of each statement and determine whether it is true or false.

Statement	**Meaning**	**Truth Value**
7. $3 < 3$		
8. $4 \leq 8$		
9. $-1 > -4$		
10. $9 = 8$		
11. $10 \neq 10$		
12. $8 \geq -10$		
13. $5 \geq 5$		
14. $-1 \leq -4$		
15. $-2 > -12$		
16. $-6 < -6$		
17. $-8 < -2$		
18. $7 \neq 9$		

Section 1.3: Operations on Real Numbers

Having defined the set of real numbers, we can use them to solve problems. But before we can do that, we have to define a set of rules that show how we can process them. In mathematics we call such a process an **operation** (a rule or set of rules) we perform on the real numbers to get a new number. In this section we define a variety of operations on real numbers that will enable us to use them to solve problems.

Objective A: Absolute Value

Sometimes we may need to consider the size of a debt or asset. With this type of problem we are not concerned with its order or precise location on the number line, but only how far it is away from zero. Thus, the sign of the number is not important, only its magnitude, i.e. how far it is away from zero.

The distance finding operator is given the name of the **absolute value** and it is calculated as follows. If the number is positive, we do nothing and just return the positive number. If, however, the number is negative, we return the opposite of the negative number. The opposite of a number is the number on the other side of 0 that has the same distance from 0 as the given number.

We should point out that distance (a measure of how far away something is) is always positive. We would never say that we are negative four feet away from an object. We say that we are four feet away from it. Thus, for example the numbers 4 and –4 are both four units away from zero as shown in the figure below. The number –4 is four units to the left of 0 and 4 is four units to the right of 0. The sign of a number tells us its direction (location) relative to zero.

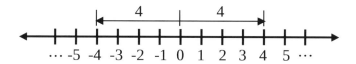

We represent the absolute value operator as two vertical lines surrounding the number, |number|. Thus, when we see |-2|, we recognize that we are finding the distance the number -2 is from zero and, since the number is negative, we change the sign and the answer is 2.

Putting this all together we can write the following formal (mathematical) definition of the absolute value operator, operating on any number x, where x can be any (positive or negative) real number.

Definition: The **absolute value** of a number is its distance from zero.

$$|x| = \begin{cases} x & \text{if } x \geq 0 \\ -x & \text{if } x < 0 \end{cases}$$

The distance of number from zero is equal to the number itself if the number is positive or zero. If the number is negative, we use its opposite of the number to indicate its distance from zero.

Example 1: Evaluate each of the following absolute value expressions.
  a) $|3|$
  b) $|-13.9|$
  c) $|0|$
  d) $|-\pi|$

Solution: a) $|3| = 3$, since 3 is not negative we leave the sign the same.

  b) $|-13.9| = -(-13.9) = 13.9$, since -13.9 is negative we change the sign to positive.

  c) $|0| = 0$, since 0 is neither negative nor positive we leave it as it is.

  d) $|-\pi| = -(-\pi) = \pi$, since $-\pi$ is negative we change the sign to positive.

ABS Operator

The **ABS** operator is used to find the absolute value of a number when we write a **BASIC** program. Try the following examples on the computer.

Example 2: What will be the output of each of the following statements?
  a) PRINT ABS(4.7)
  b) PRINT ABS(-9)
  c) PRINT ABS(-5.23)

Solution: a) The output of the statement is 4.7.
  b) The output of the statement is 9.
  c) The output of the statement is 5.23.

The absolute value operator will be used later in more complicated expressions. We will now review the rules for addition, subtraction, multiplication, and division of real numbers.

Objective B: Addition and Subtraction of Real Numbers

Now that we have defined our set of real numbers and can talk about their order relative to one another and how far they are away from zero on the number line, we can introduce more complicated operations on these numbers. We begin with the operations of basic arithmetic. We start by defining the various operators using natural numbers and then extend the definition to all real numbers. It is easier to comprehend with natural numbers, but the rules we show for the natural numbers apply equally to all real numbers.

Adding Real Numbers

Adding two numbers means we are finding the final location on the number line after performing our addition operation on the two numbers. For example, to add two numbers 2 and 3 is equivalent to first moving 2 units from zero to the right on the number line and then moving an additional 3 units to the right. This takes us to the final location of 5 units to the right of zero on the number line.

Or similarly, if we are given 2 dollars and then given an additional 3 dollars we would have a total of 5 dollars.

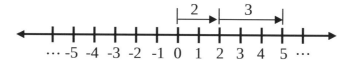

Mathematically we express this using the addition operator, +, and write this as 2 + 3 = 5.

Now, add the numbers –2 and –3. The –2 takes us two units to the left of zero and then the –3 takes us an additional three units further to the left (we moved left since both of the numbers were negative). This takes us to the final location of 5 units to the left of zero on the number line.

We can also think of this using the following equivalent representation; if we are 2 dollars in debt (-2) and then add another debt of 3 dollars (-3), we are now in debt 5 dollars (-5).

Mathematically we express this as: (-2) + (-3) = (-5).

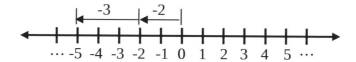

To add two numbers of different signs, we follow the same procedure as above. For example, adding -3 and +4 together, we first move three units to the left of zero putting us at the −3 location on the number line, and then move four units to the right (since the 4 is positive). This puts us at the final location of a positive one (1) on the number line. This is equivalent to being in debt 3 dollars (-3) and then getting 4 dollars (4) to leave us with a net gain of 1 dollar (1).

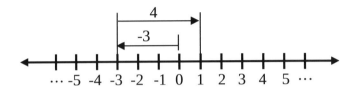

Mathematically we write $-3+4=1$.

We summarize the rules for adding two real numbers as follows:

Like Signs. Add two numbers with the same sign by adding their absolute values and taking the sign of the original numbers.

Unlike Signs. Add two numbers with different signs by subtracting the smaller absolute value from the larger and taking the sign of the number with the larger absolute value.

Example 1: Adding numbers with like signs: a) $5+7$ b) $(-5)+(-7)$

  Solution: a) $5+7=+(|5|+|7|)=+(5+7)=+12$

  b) $(-5)+(-7)=-(|-5|+|-7|)=-(5+7)=-12$

Example 2: Adding numbers with unlike signs: a) $+5+(-7)$ b) $(-5)+(+7)$

  Solution: a) $+5+(-7)=-(|-7|-|+5|)=-(7-5)=-2$

  b) $-5+(+7)=+(|+7|-|-5|)=+(7-5)=+2$

Example 3: Add: a) $8+(-6)$ b) $-25+(-23)$ c) $-4.5+2.3$

  Solution: a) $8+(-6)=2$. Since the two numbers have unlike signs, subtract the smaller absolute value from the larger and keep the sign of the number with the larger absolute value.

b) $-25+(-23)=-48$ Since the two numbers have like signs, add their absolute values and keep the negative sign.

c) $-4.5+2.3=-2.2$

Example 4: Add: $-\dfrac{3}{5}+\left(-\dfrac{4}{5}\right)$

Solution: $-\dfrac{3}{5}+\left(-\dfrac{4}{5}\right)=-\dfrac{7}{5}$ NOTE: Remember to find the least common denominator if they are different.

Subtracting Real Numbers

Subtracting (or finding the difference) of any two real numbers is expressed mathematically by using the subtraction operator, $-$. Subtraction can be viewed as "adding the opposite."

Subtraction is very similar to doing the opposite of addition. In fact, one approach is to transform our subtraction problem into addition and then perform the addition operation.

To change subtraction into addition we simply change the sign of the number we are subtracting and then add the resulting two numbers together. Remember that the opposite of a number has the same distance from zero but in the opposite direction. Hence the opposite of 2 is –2 and the opposite of –5 is +5. We will discuss this more in the next section.

Thus,
$8-3$ becomes: $8+(-3)=5$

and
$(-7)-(5)$ becomes: $(-7)+(-5)=-12$.

In other words, if we initially have 8 dollars and subtract 3 dollars from it, it is equivalent to adding a debt of 3 dollars to the initial 8 dollars giving us a net of 5 dollars. Or in the second example, if we are in debt 7 dollars and subtract 5, it is equivalent to being in debt 7 dollars and adding an additional debt of 5 dollars, resulting in a debt of 12 dollars

Example 5: Subtract : a) $7-(-8)$ b) $2-6$ c) $-6-(-9)$ d) $-4.5-2.3$

Solution: a) $7-(-8) = 7+(+8) = 15$

b) $2-6 = 2+(-6) = -4$

c) $-6-(-9) = -6+(9) = 3$

d) $-4.5-2.3 = -4.5+(-2.3) = -6.8$

Example 6: Subtract : a) $\dfrac{3}{5} - \dfrac{4}{5}$ b) $\dfrac{-2}{9} - \left(\dfrac{-2}{3}\right)$

Solution: a) $\dfrac{3}{5} - \dfrac{4}{5} = \dfrac{3}{5} + \left(\dfrac{-4}{5}\right) = -\dfrac{1}{5}$

b) $-\dfrac{2}{9} - \left(\dfrac{-2}{3}\right) = -\dfrac{2}{9} - \left(\dfrac{-6}{9}\right) = -\dfrac{2}{9} + \dfrac{6}{9} = \dfrac{-2+6}{9} = \dfrac{4}{9}$

NOTE: The least common denominator in this problem is 9.

Objective C: Inverse Operations and Inverse Operators

Addition and subtraction have a special relationship to one another. They are called inverse operations of each other. An inverse operation is an operation that undoes what the other operator has done. For example, if we add 5 to a number we can undo this by subtracting 5 from the result, i.e. we start with 8 and add 5 to obtain $8 + 5 = 13$ and then subtracting 5 gets us back to 8, that is, $13-5=8$.

Or, alternatively, we could consider the "operation" of tying your shoe. Obviously the inverse operation is untying the shoe. We say that tying and untying are inverse operations of one another, just as addition and subtraction are.

An inverse operator gets you back where you started. In addition, we can also call the inverse of a number the opposite of the number, or the additive inverse. For example, -5 is the inverse or opposite of +5.

It is very important to understand this seemingly abstract concept of an operator and its inverse operator when working with computers.

Example : Find the additive inverse of each of the following numbers.

Number	Additive inverse
5	–5
–7	7
$\frac{3}{4}$	$-\frac{3}{4}$
$-\frac{1}{5}$	$\frac{1}{5}$
2.9	–2.9
–3.6	3.6
$|-3|$	–3
$-|-3|$	3
–(–5)	–5

Objective D: Multiplication and Division of Real Numbers

Multiplying Real Numbers

Multiplying or finding the product of two real numbers can be interpreted as performing addition a "multiple" number of times. The multiplication operator is represented by one of three different symbols, either by placing ×, * or · between the two numbers. We will not use the × symbol. Instead, we will use the symbols * or · interchangeably in this text. On a computer we reserve the asterisk, *, to represent multiplication. The other symbols and conventions are not recognized and should not be used in programs.

Suppose we want to multiply 4 by 3, that would be equivalent to taking the number 3 and adding it to itself 4 times 3 + 3 + 3 + 3, which is equal to 12. Written in operator notation we could either write this as:

4*3=12,   4·3=12   or   3(4)=12   (This is assumed to be multiplication.)

Or if we want the product of –3 and 4 we would have –3 added to itself, 4 times, producing a –12.

The following rules apply when multiplying any real numbers:

Like Signs: Multiply two numbers by multiplying their absolute values and taking the sign to be positive.

Unlike Signs: Multiply the two numbers by taking the product of their absolute values and taking the sign to be negative.

Example 1: Multiply: a) $-4*(6)$ b) $(-2)*(-5)$ c) $4.3*(-3)$ d) $-2*\left(\dfrac{3}{4}\right)$
e) $(2.9)*(3.4)$ f) $(-5)*(4.1)$

Solution: a) $-4*(6) = -24$

b) $(-2)*(-5) = 10$

c) $4.3*(-3) = -12.9$

d) $-2*\left(\dfrac{3}{4}\right) = -\dfrac{3}{2}$

e) $(2.9)*(3.4) = 9.86$

f) $(-5)*(4.1) = -20.5$

Dividing Real Numbers

Dividing or finding the quotient of two real numbers is represented mathematically with one of two symbols; /, or ÷. Thus, 15 divided by 3 can be written as $15/3$ or $\dfrac{15}{3}$ or $15 \div 3$.

On a computer we reserve the forward slash, /, to represent division. So we type 15/3. Just as subtraction can be viewed in terms of addition, division can be viewed in terms of multiplication. Division is the same as multiplying by the reciprocal of the second number. Thus,

$$3 \div 15 = 3 * \dfrac{1}{15} = \dfrac{3}{1} * \dfrac{1}{15} = \dfrac{3*1}{1*15} = \dfrac{3}{15} = \dfrac{\cancel{3}}{\cancel{3}*5} = \dfrac{1}{5}$$

and:

$$\frac{2}{3} \div \frac{5}{2} = \frac{2}{3} * \frac{2}{5} = \frac{2*2}{3*5} = \frac{4}{15}$$

Also, we can make the following observations as we did with multiplication above:

Like Signs: Divide two numbers by dividing their absolute values and taking the sign to be positive.

Unlike Signs: Divide the two numbers by dividing their absolute values and taking the sign to be negative

We have one caveat when we divide by real numbers; that is, we can never divide by zero! Anything divided by zero is undefined.

Example 2: Divide: a) $\frac{-10}{-5}$ b) $\frac{-15}{3}$ c) $\frac{2}{3} \div \frac{4}{5}$ d) $\left(-\frac{3}{8}\right) \div \left(\frac{3}{4}\right)$ e) $4 \div \left(-\frac{2}{5}\right)$

Solution: a) $\frac{-10}{-5} = \frac{\cancel{2}*\cancel{5}}{\cancel{5}} = 2$

b) $\frac{-15}{3} = \frac{-\cancel{3}*5}{\cancel{3}} = -5$

c) $\frac{2}{3} \div \frac{4}{5} = \frac{2}{3} * \frac{5}{4} = \frac{2*5}{3*4} = \frac{\cancel{2}*5}{3*2*\cancel{2}} = \frac{5}{6}$

d) $\left(-\frac{3}{8}\right) \div \left(\frac{3}{4}\right) = -\frac{3}{8} * \frac{4}{3} = \frac{-3*4}{3*8} = \frac{-\cancel{3}*\cancel{4}}{\cancel{3}*2*\cancel{4}} = -\frac{1}{2}$

e) $4 \div \left(-\frac{2}{5}\right) = \frac{4}{1} \div \left(-\frac{2}{5}\right) = \frac{4}{1} * \left(-\frac{5}{2}\right) = \frac{4*(-5)}{1*2} = \frac{-20}{2} = -10$

Example 3: Divide: $\dfrac{-15}{0}$

Solution: We cannot divide by zero, this expression is undefined.

NOTE: A computer or calculator may return an "error" message if you try this.

Just as addition and subtraction were inverse operators, multiplication and division are inverse operators of each other. For example; $7 \cdot 3 = 21$ and then $21 \div 3 = 7$ so multiplying by 3 then dividing by 3 gets us back to where we started.

Objective E: Exponents of Real Numbers

We will often have to multiply a number by itself several times, such as 3 multiplied by itself 5 times. Instead of writing 3*3*3*3*3 all the time, we define a new operator called the exponent or power operator.

Thus, instead of 3*3*3*3*3 we would write 3^5 in a mathematical expression or 3^5 when using a computer. The symbol "^" is called a carat and is used for describing exponents. The exponent operator is simply a shorthand way of writing a number multiplied by itself a fixed number of times. The number we are multiplying is called the base of the exponent operator and the number of times we are multiplying the number by itself is called the exponent.

Examples:

1. $2 \wedge 4 = 2^4 = 2*2*2*2 = 16$, the base is 2 and the exponent is 4, read as "two to the fourth" or "two to the fourth power"

2. $3 \wedge 6 = 3^6 = 3*3*3*3*3*3 = 729$, the base is 3 and the exponent is 6, read as "three to the sixth" or "three to the sixth power"

3. $5 \wedge 3 = 5^3 = 5*5*5 = 125$, the base is 5 and the exponent is 3, read as "five to the third" or "five cubed"

4. $8 \wedge 2 = 8^2 = 8*8 = 64$, the base is 8 and the exponent is 2, read as "eight to the second" or "eight squared"

Having defined the exponent operator, some basic rules naturally follow from this definition. Also, we shall use the * symbol for multiplication in the following examples.

Rules of Exponents:

Consider what happens when we multiply two exponential terms with the same base,

Examples:

$$3^3 * 3^4 = (3*3*3)*(3*3*3*3) = 3^{3+4} = 3^7 = 2,187$$
$$2^5 * 2^4 = (2*2*2*2*2)*(2*2*2*2) = 2^{5+4} = 2^9 = 512$$
$$5^3 * 5^4 = (5*5*5)*(5*5*5*5) = 5^{3+4} = 5^7 = 78,125$$

Looking closely, we can see a pattern. In each case, we add the exponents.

> **Rule 1:** When multiplying two exponent terms with the same base, **b**, and exponents **n** and **m** we simply add the exponents:
>
> $$b^n * b^m = b^{(n+m)}$$

NOTE: We cannot combine the exponents of terms with different bases, e.g. $2^4*3^3=2*2*2*2*3*3*3$. We cannot rewrite this in terms of a single base raised to a single exponent.

Now, Consider what happens when we divide two exponent terms, again with the same base,

Examples:

$$\frac{3^5}{3^3} = \frac{(3*3*\cancel{3}*\cancel{3}*\cancel{3})}{(\cancel{3}*\cancel{3}*\cancel{3})} = 3^{(5-3)} = 3^2 = 9$$

$$\frac{2^8}{2^2} = \frac{(2*2*2*2*2*2*\cancel{2}*\cancel{2})}{(\cancel{2}*\cancel{2})} = 2^{(8-2)} = 2^6 = 64$$

$$\frac{5^7}{5^6} = \frac{(5*\cancel{5}*\cancel{5}*\cancel{5}*\cancel{5}*\cancel{5}*\cancel{5})}{(\cancel{5}*\cancel{5}*\cancel{5}*\cancel{5}*\cancel{5}*\cancel{5})} = 5^{(7-6)} = 5^1 = 5$$

We have drawn a line through the factors in the numerator and denominator that cancel out.

Another pattern is evident. We subtract the exponents.

> **Rule 2:** When dividing two exponent terms with the same base, **b**, not equal to zero, and exponents **n** and **m** we subtract the exponents:
>
> $$\frac{b^n}{b^m} = b^{(n-m)}$$

What happens when we raise an exponent term to a power,

Examples:

$$(3^2)^3 = (3^2)*(3^2)*(3^2) = (3*3)*(3*3)*(3*3) = 3^6 = 729$$

$$(2^4)^2 = (2^4)*(2^4) = (2*2*2*2)*(2*2*2*2) = 2^8 = 256$$

$$(2^3)^5 = (2^3)*(2^3)*(2^3)*(2^3)*(2^3)$$
$$= (2*2*2)*(2*2*2)*(2*2*2)*(2*2*2)*(2*2*2)$$
$$= 2^{3*5} = 2^{15} = 32,768$$

> **Rule 3:** When raising an exponent term to a power, the general rule is that for any base, **b**, and exponents, **n** and **m** we multiply the exponents:
>
> $$(b^n)^m = b^{n*m}$$

Now let us see what happens when we divide an exponent term by the same exponent term and apply **Rule 2** above,

Examples:

$$\frac{3^3}{3^3} = \frac{(\cancel{3}*\cancel{3}*\cancel{3})}{(\cancel{3}*\cancel{3}*\cancel{3})} = 3^{(3-3)} = 3^0 = 1$$

$$\frac{2^6}{2^6} = \frac{(\cancel{2}*\cancel{2}*\cancel{2}*\cancel{2}*\cancel{2}*\cancel{2})}{(\cancel{2}*\cancel{2}*\cancel{2}*\cancel{2}*\cancel{2}*\cancel{2})} = 2^{(6-6)} = 2^0 = 1$$

$$\frac{5^4}{5^4} = \frac{(\not{5}*\not{5}*\not{5}*\not{5})}{(\not{5}*\not{5}*\not{5}*\not{5})} = 5^{(4-4)} = 5^0 = 1$$

The pattern that emerges here is for any base **b** not equal to zero:

Rule 4: Any nonzero quantity raised to the zero power is always one.
$$b^0 = 1$$

Suppose we want to raise the product of two numbers to an exponent,

Examples:
$$(2*4)^3 = (2*4)*(2*4)*(2*4) = 2^3 * 4^3 = 8*64 = 512$$

$$(11*3)^2 = (11*3)*(11*3) = 11^2 * 3^2 = 121*9 = 1,089$$

Rule 5: when raising the product of two non-zero bases to an exponent **n**:
$$(ab)^n = a^n b^n$$

Now, let us consider what happens when we raise a fractional term to a power,

Examples:

$$\left(\frac{3}{4}\right)^3 = \left(\frac{3}{4}\right)*\left(\frac{3}{4}\right)*\left(\frac{3}{4}\right) = \frac{(3*3*3)}{(4*4*4)} = \frac{3^3}{4^3} = \frac{27}{64}$$

$$\left(\frac{5}{7}\right)^2 = \left(\frac{5}{7}\right)*\left(\frac{5}{7}\right) = \frac{(5*5)}{(7*7)} = \frac{5^2}{7^2} = \frac{25}{49}$$

$$\left(\frac{6}{11}\right)^5 = \left(\frac{6}{11}\right)*\left(\frac{6}{11}\right)*\left(\frac{6}{11}\right)*\left(\frac{6}{11}\right)*\left(\frac{6}{11}\right)$$
$$= \frac{(6*6*6*6*6)}{(11*11*11*11*11)} = \frac{6^5}{11^5} = \frac{7,776}{161,051}$$

> **Rule 6:** For any fractional base, $\dfrac{a}{b}$, and exponent **n**:
> $$\left(\dfrac{a}{b}\right)^n = \dfrac{a^n}{b^n}$$

Now, let us look at what happens when we divide an exponent term by another exponent term to the same base, but raised to a higher power, and apply **Rule 2** from above,

Examples:

$$\dfrac{3^2}{3^4} = \dfrac{(\cancel{3}*\cancel{3})}{(3*3*\cancel{3}*\cancel{3})} = \dfrac{1}{(3*3)} = \dfrac{1}{3^2} = 3^{(2-4)} = 3^{-2}$$

$$\dfrac{2^2}{2^6} = \dfrac{(\cancel{2}*\cancel{2})}{(2*2*2*2*\cancel{2}*\cancel{2})} = \dfrac{1}{(2*2*2*2)} = \dfrac{1}{2^4} = 2^{(2-6)} = 2^{-4}$$

$$\dfrac{5}{5^4} = \dfrac{(\cancel{5})}{(5*5*5*\cancel{5})} = \dfrac{1}{(5*5*5)} = \dfrac{1}{5^3} = 5^{(1-4)} = 5^{-3}$$

$$\dfrac{5^0}{5^4} = \dfrac{1}{(5*5*5*5)} = \dfrac{1}{5^4} = 5^{(0-4)} = 5^{-4}$$

Now we notice that in this case the final exponent is negative and that the following is true:

> **Rule 7:** Any non-zero base raised to a negative exponent can be written as
> $$b^{-n} = \dfrac{1}{b^n} \quad \text{or} \quad \dfrac{1}{b^n} = b^{-n}$$

The above rules also apply to exponential expressions whose bases and exponents are rational and irrational numbers, although we will not show it here.

Summary of the Rules of Exponents:
let "a" and "b" be non-zero real numbers, "n" and "m" be any real numbers, then

    1. $b^n * b^m = b^{(n+m)}$

2. $\dfrac{b^n}{b^m} = b^{(n-m)}$

3. $b^0 = 1$

4. $(b^n)^m = b^{n*m}$

5. $(ab)^n = a^n b^n$

6. $\left(\dfrac{a}{b}\right)^n = \dfrac{a^n}{b^n}$

7. $\dfrac{1}{b^n} = b^{-n}$ or $b^{-n} = \dfrac{1}{b^n}$

Examples: Evaluate each of the following using the rules of exponents:

1. $5^3 * 5^4 = 5^{(3+4)} = 5^7 = 78,125$ We add the exponents.

2. $\dfrac{8^5}{8^3} = 8^{(5-3)} = 8^2 = 64$ We subtract the exponents.

3. $\left(4^2 * 5^3\right)^0 = 1$ A quantity raised to zero equals 1.

4. $\left(\dfrac{4}{7}\right)^3 * \left(\dfrac{4}{7}\right)^7 = \left(\dfrac{4}{7}\right)^{10} = \dfrac{4^{10}}{7^{10}} = \dfrac{1,048,576}{282,474,249}$

5. $\dfrac{12^3}{12^6} = 12^{(3-6)} = 12^{-3} = \dfrac{1}{12^3} = \dfrac{1}{1728}$

6. $\dfrac{3^4 * 5^2}{3^2 * 5^4} = 3^{(4-2)} * 5^{(2-4)} = 3^2 * 5^{-2} = \dfrac{3^2}{5^2} = \dfrac{9}{25}$

7. $2^3 * 3^4 * 2^2 * 3^5 = 2^{(3+2)} * 3^{(4+5)} = 2^5 * 3^9 = 32 * 19,683 = 629,856$

8. $\left(4^2\right)^3 = 4^6 = 4,096$

9. $\left(\left(\dfrac{4}{5}\right)^3\right)^2 = \left(\dfrac{4}{5}\right)^{3 \cdot 2} = \left(\dfrac{4}{5}\right)^6 = \dfrac{4^6}{5^6} = \dfrac{4,096}{15,625}$

10. $\dfrac{10^3}{10^6} = 10^{-3} = \dfrac{1}{10^3} = \dfrac{1}{1,000}$

11. $-(-3)^2 = -(-3)*(-3) = -9$ Perform $(-3)^2$ first, then find its opposite.
12. $(-|-5|)^2 = (-5)^2 = 25$
13. $-3^2 + (-3)^2 = -3*3 + (-3)*(-3) = -9 + 9 = 0$

EXERCISES 1.3

Find the additive inverse of each of the following numbers.
1. 5
2. −7
3. $\dfrac{3}{4}$
4. $-\dfrac{1}{5}$
5. 2.9
6. −3.6

Evaluate each of the following:
7. $|-5|$
8. $|9|$
9. $|1.23|$
10. $|-56|$
11. $|-7|$
12. $|-4.7|$
13. $\left|\dfrac{3}{8}\right|$
14. $\left|-\dfrac{1}{3}\right|$
15. $|0.45|$
16. $|-8.4|$
17. $3+5$
18. $-8+(-2)$
19. $7+(-1)$
20. $-5+10$
21. $-10+(-7)$
22. $(-6)+(-9)$
23. $-\dfrac{1}{5}+\dfrac{3}{5}$
24. $\dfrac{3}{7}+\left(-\dfrac{2}{7}\right)$
25. $-\dfrac{3}{8}+\dfrac{5}{12}$
26. $-\dfrac{3}{5}+\left(\dfrac{3}{4}\right)$
27. $1.2+(-4)$
28. $(-4.3)+(-1.6)$
29. $-4.5+7.4$

30. $-3.45+3.45$

31. $0+(-21.7)$

32. $7-4$

33. $5-9$

34. $-4-3$

35. $-7-(-3)$

36. $-5-8$

37. $-10-(-23)$

38. $18-20$

39. $11-(-6)$

40. $-\dfrac{2}{7}-\dfrac{3}{7}$

41. $\dfrac{3}{5}-\dfrac{4}{5}$

42. $-\dfrac{2}{9}-\left(\dfrac{-2}{3}\right)$

43. $\dfrac{7}{12}-\left(-\dfrac{5}{18}\right)$

44. $-4.5-2.3$

45. $5.7-7.5$

46. $3.8-(-1.4)$

47. $-4*(6)$

48. $(-2)\cdot(-5)$

49. $4.3*(-3)$

50. $-2*\left(\dfrac{3}{4}\right)$

51. $(2.9)\cdot(3.4)$

52. $(-5)\cdot(4.1)$

53. $\left(\dfrac{5}{14}\right)*\left(-\dfrac{7}{10}\right)$

54. $\dfrac{-10}{-5}$

55. $\dfrac{-15}{3}$

56. $\dfrac{24}{8}$

57. $\dfrac{36}{-9}$

58. $\dfrac{2}{3}\div\dfrac{4}{5}$

59. $\left(-\dfrac{3}{8}\right)\div\left(\dfrac{3}{4}\right)$

60. $4\div\left(-\dfrac{2}{5}\right)$

Evaluate using the rules of exponents.

61. $(-2)^4$

62. 5^2

63. $\left(\dfrac{1}{6}\right)^2$

64. $(-2)^3$

65. $\left(-\dfrac{4}{7}\right)^2$

66. $\left(\dfrac{15}{9}\right)^1$

67. $(-5)^0$

68. 3^2*3^4

69. $(-4)^2(-4)^3$

70. $(7*3)^4$

71. $\dfrac{4^5}{4^3}$

72. $\dfrac{(-5)^7}{(-5)^4}$

73. 7^{-2}

74. $\left(\dfrac{2}{3}\right)^{-3}$

75. $\left(\dfrac{21}{100}\right)^0$

76. 3^{-4}

77. $\dfrac{1}{5^{-2}}$

78. $(-31)^{-1}$

79. $(-4)^{-3}$

80. $\dfrac{5^{-4}}{5^{-6}}$

COMPUTER ACTIVITY

Write a program that performs the following operations. Compare your output with the answers you obtained when doing the calculations yourself.

1. $|-8.4|$
2. $(-4.3)+(-1.6)$
3. $-\left(\dfrac{4}{7}\right)^2$
4. $(-4)^{-3}$
5. $(-4)^2(-4)^3$
6. $(-2)^4$

Section 1.4 Additional Operations and Properties of Real Numbers

In this section, we are going to discuss roots and radicals, percents, order of operations, and properties of real numbers. We shall also write BASIC programs to demonstrate these mathematical concepts.

Objective A: Roots and Radicals of Real Numbers

Just as addition and multiplication had inverse operations, the same is true for the exponent (power) operator. The inverse operator of an exponent is called a radical or root operator.

For example, consider the exponent operator defined as the square of a number, i.e.

$$3 \wedge 2 = 3^2 = 9$$

We started with the base 3 and squared it (exponent of 2) to get the number 9.

The inverse operation starts with the number 9 and gets us back to the base of 3. The inverse operator of the squaring operator is the square root or radical operator. It undoes the squaring operation. More specifically, the square root operator asks the question: what number (base) multiplied by itself 2 (exponent) times gives you 9? The answer is 3.

Mathematically we write this as

$$\sqrt{9} = 3$$

The symbol, $\sqrt{}$, is called a radical symbol and the number inside the radical symbol is called the radicand.

We should, however, point out that the number 9 has two square roots. One is 3 and the other one is -3 since 3^2 and $(-3)^2$ are both equal to 9. We write the positive square root of 9 as $\sqrt{9} = 3$. We call this the principal square root of 9. We write the negative square root of 9 as $-\sqrt{9} = -3$.

We should also point out that the inverse operation is not true for all real numbers. In particular, $\sqrt{-9}$ is undefined. This is because, by our definition, there is no real number when multiplied by itself will give us a -9. In fact, no real number when multiplied by itself can equal a negative number. Thus, the square root of any negative number is undefined.

NOTE: Some of you may have studied imaginary numbers, so you know that $\sqrt{-9} = 3i$, but in this course we are only concerned with real numbers so we will say that $\sqrt{-9}$ is undefined.

Associated with every exponent, there is a root (radical) that will undo it. Here are a few of the more common ones. For simplicity we only focus on integer exponents. What we show here also applies to any rational exponent, but we will not go into that level of detail here.

Example 1: Evaluate a) $\sqrt{25}$ b) $-\sqrt{81}$ c) $\sqrt{\dfrac{16}{49}}$ d) $\sqrt{0.09}$

Solution: a) $\sqrt{25} = 5$ since $5*5 = 25$

b) $-\sqrt{81} = -9$ since $9*9 = 81$

c) $\sqrt{\dfrac{16}{49}} = \dfrac{\sqrt{16}}{\sqrt{49}} = \dfrac{4}{7}$ since $4*4 = 16$ and $7*7 = 49$

d) $\sqrt{0.09} = 0.3$ since $0.3*0.3 = 0.09$

Example 2: Evaluate $\sqrt{7}$ Round your answer to the nearest hundredths.

Solution: $\sqrt{7} = 2.65$ NOTE: 7 is not a perfect square.

Example 3: Evaluate $-\sqrt{19}$ Round your answer to the nearest tenths.

Solution: $-\sqrt{19} = -4.4$

SQR Operator in BASIC

The SQR operator is used to find the square root of a number whenever we write a BASIC program. SQR returns the positive square root, therefore,

$$\sqrt{x} = \text{SQR}(x).$$

Example 4: What will be the output of each of the following?
  a) `PRINT   SQR(4)`

b) PRINT   SQR(9+16)
c) PRINT   SQR(2^5 + 4^2 + 1)

Solution: a) The output should be 2.
b) The output should be 5.
c) The output should be 7.

Example 5:   What message will you see when you enter and run this line?

PRINT   SQR(-9)

Perhaps some of you have studied rational exponents in another course so you know $x^{1/2} = \sqrt{x}$, $x^{1/3} = \sqrt[3]{x}$, etc.... Instead of using the SQR operator to find the square root of a number, we can also use the ^ symbol and 1/2 to take the square root of a number in writing a program. That is, $\sqrt{4} = 4 \wedge (1/2)$ is equal to 2.

The cube root of a number, $\sqrt[3]{}$, is a number that if it is multiplied by itself three times, it gives you the radicand or the number inside the radical symbol.

Examples:

$$\sqrt[3]{8} = 2 \quad \text{since } 2*2*2 = 8$$

$$\sqrt[3]{-27} = -3 \text{ since } (-3)*(-3)*(-3) = -27$$

In BASIC programming, there is no built-in operator that we can use to take the cube root of a number. So we use ^(1/3) to do it. For the example we have above, $\sqrt[3]{8} = 8 \wedge (1/3) = 2$ and $\sqrt[3]{-27} = (-27) \wedge (1/3) = -3$. We would write the code as follows:

PRINT 8^(1/3)   The output for this should be 2.

PRINT (-27)^(1/3)   The output for this should be -3.

Objective B: Properties of Real Numbers

The arithmetic operators addition and multiplication are governed by the following laws:

Associative Laws

When adding three numbers, it makes no difference which where you locate the parenthesis. We could add the first and second numbers first or we could add the second and third numbers first. This is called the associative law of addition.

$$(a+b)+c = a+(b+c)$$

Example:

$(2+5)+7 = 7+7 = 14$
$2+(5+7) = 2+12 = 14$ so $(2+5)+7 = 2+(5+7) = 14$

Caution: The associative law is **not** true for subtraction:

$$(a-b)-c \neq a-(b-c)$$

To illustrate this:

$(2-5)-7 = -3-7 = -10$
$2-(5-7) = 2-(-2) = 4$ so $(2-5)-7 \neq 2-(5-7)$

However, if we convert the subtraction to addition and then apply the Associative law

$(2+(-5))+(-7) = (-3)+(-7) = -10$
$2+(-5+(-7)) = 2+(-12) = -10$ so $(2+(-5))+(-7) = 2+(-5+(-7)) = -10$

The associative law is true for multiplication:

$$(a*b)*c = a*(b*c)$$

Example:

$(3*4)*5 = 12*5 = 60$
$3*(4*5) = 3*20 = 60$ so $(3*4)*5 = 3*(4*5)$

Caution: The associative law is **not** true for division:

$(a \div b) \div c \neq a \div (b \div c)$

To illustrate this:

$$(100 \div 50) \div 2 = 2 \div 2 = 1$$
$$100 \div (50 \div 2) = 100 \div 25 = 4$$
so $(100 \div 50) \div 2 \neq 100 \div (50 \div 2)$

Commutative Laws

Two numbers can be added in either order. This is called the Commutative Law of Addition.
$$a + b = b + a$$

Example:
$$2 + 5 = 7$$
$$5 + 2 = 7$$
so $2 + 5 = 5 + 2 = 7$

Two numbers can be multiplied in either order. This is called the Commutative Law of Multiplication.
$$a * b = b * a$$

Example:
$$5 * 2 = 10$$
$$2 * 5 = 10$$
so $5 * 2 = 2 * 5 = 10$

Once again, we should point out that the commutative laws do not work for subtraction or division.

Identity Laws

For addition, the identity is the number that we add to a given number *a* so that the result is the number *a* itself. The identity for addition is **0**.

$$a + 0 = 0 + a = a$$

Example:
$$5 + \mathbf{0} = \mathbf{0} + 5 = 5$$

We call **0** the additive identity.

For multiplication, the identity is the number that we multiply with a number *a* so that the result is the number *a* itself. The identity for multiplication is **1**.

$$a * 1 = 1 * a = a$$

Example:
$$5 * 1 = 1 * 5 = 5$$

We call the number **1** the multiplicative identity.

Inverse Laws

For addition, the inverse of a number *a* is that number that is added to *a* so that the result is 0. This number is the opposite of *a*.

$$a+(-a)=(-a)+a=0$$

Example:
$$5+(-5)=(-5)+5=0$$

We call –*a* the additive inverse *a* or simply the opposite of *a*.

For multiplication, the inverse of a number *a* is that number multiplied with *a* so that the result is 1. The exception is that *a* cannot be zero. This number is the reciprocal of *a*.

$$a*\left(\frac{1}{a}\right) = \left(\frac{1}{a}\right)*a = 1, \ a \neq 0.$$

Example:
$$5*\left(\frac{1}{5}\right) = \left(\frac{1}{5}\right)*5 = 1$$

We call $\frac{1}{a}$ the multiplicative inverse of *a* or the reciprocal of *a*, provided *a* is not 0.

Distributive Law

Multiplying the sum of two or more terms by some number *n*, is the same as the sum of the individual products.

$$n*(a+b+c)=n*a+n*b+n*c \quad \text{or}$$

$$(a+b+c)*n=a*n+b*n+c*n$$

This is called the distributive law of multiplication over addition.

Example:

$(3+5+7)*2 = (15)*2 = 30$
$3*2+5*2+7*2 = 6+10+14 = 30$

so $(3+5+7)*2 = 3*2+5*2+7*2 = 30$

The properties of real numbers are summarized in the following table:

For real numbers a, b, c, and n,

Property	Addition	Multiplication
Commutative Law	$a+b=b+a$	$a*b=b*a$
Associative Law	$(a+b)+c = a+(b+c)$	$(a*b)*c = a*(b*c)$
Identity Law	$a+(0) = (0)+a = a$	$a*1 = 1*a = a$
Inverse Law	$a+(-a) = (-a)+a = 0$	$a*\left(\frac{1}{a}\right) = \left(\frac{1}{a}\right)*a = 1, a \neq 0$
Distributive Law of Multiplication over Addition	$n*(a+b+c) = n*a+n*b+n*c$	

Examples: *State the property or properties used in each of the following:*

Exercise	Property
1. $(2+3)+4 = 2+(3+4)$	Associative law of addition
2. $(9*3)*4 = 9*(3*4)$	Associative law of multiplication
3. $4*(3+7+9) = 4*3+4*7+4*9$	Distributive law of multiplication over addition
4. $55+74 = 74+55$	Commutative law of addition
5. $12*55 = 55*12$	Commutative law of multiplication
6. $99+34 = 34+99$	Commutative law of addition
7. $121+0 = 121$	Identity for addition
8. $(34+15)*3 = 34*3+15*3$	Distributive law of multiplication over addition

Objective C: Percents

When we buy an item from the department store, we have to pay 8.75% sales tax. We take a loan to buy a car and have a pay an interest rate of 7%. The price of gasoline has gone up from $1.69 to $2.29 in one month; what is the percent increase in price? What do we mean by percent?

The word **percent** means out of a hundred or per hundred. The symbol % means $\frac{1}{100}$ or 0.01. When we write 50%, we mean $50\left(\frac{1}{100}\right) = \frac{50}{100} = \frac{1}{2}$ or we can write 50% as $50(0.01) = 0.5$. Most of the time when we work with percent, we need to convert the percent to decimal.

To convert a percent to decimal, move the decimal point two places to the left and drop the percent symbol inserting zeros, if necessary. For example to convert 8.75% to decimal, moving the decimal point which is after the 8 two places to the left and drop the % symbol, we get 0.0875. Notice that we inserted a 0 before 8 as a placeholder.

To convert a decimal to percent, move the decimal point two places to the right and attach the percent symbol inserting zeros, if necessary. For example, to convert 0.35 to percent, move the decimal point before 3 two places to the right placing the decimal point after 5, then attach the percent symbol. We get 35%.

Knowing the percent equation: amount = percent * base and basic algebra, we can solve many problems that deal with percents. We shall show how to solve each of the following types of percent problems.

Example 1:   What is 15% of 200?

   Solution: Here 15 is the percent and the base is 200. So we are finding the amount.

   amount = 15% * 200   changing 15% to decimal
   = 0.15 * 200
   = 30

   30 is 15% of 200.

Example 2:   25 is what percent of 80?

   Solution: Here 25 is the amount and the base is 80. So we are finding the percent.

$$25 = \text{percent} * 80 \quad \text{divide both sides by 80}$$

$$\text{percent} = \frac{25}{80} \quad \text{convert } \frac{25}{80} \text{ to decimal}$$

$$= 0.3125 \quad \text{convert to percent}$$

$$= 31.25\%$$

25 is 31.25% of 80.

Example 3:   35 is 25% of what number?

Solution: Here 35 is the amount and the percent is 25. So we are finding the base.

$$35 = 25\% * \text{base} \quad \text{convert 25\% to decimal}$$

$$35 = 0.25 * \text{base} \quad \text{divide both sides by 0.25}$$

$$\text{base} = \frac{35}{0.25}$$

$$= 140$$

35 is 25% of 140.

Suppose we want to find the percent increase or decrease from the original quantity to the new quantity, we subtract the smaller of the two quantities from the larger, then divide by the original quantity.

Example 4:   The price of gasoline went up from $1.69 to $2.19 in one month, find the percent increase.

Solution: percent increase $= \dfrac{2.19 - 1.69}{1.69} = \dfrac{0.50}{1.69} = 0.2959 \approx 30\%$. Notice we use the original quantity 1.69 for the divisor.

NOTE: It is important to divide the difference by the original quantity.

Objective D: Order of Operations

In section 1.3 we defined different operations and their inverses: addition and subtraction, multiplication and division, exponentiation (power) and root. We looked at each separately. However, most calculations involve many if not all of the operators in a single expression, and this can cause confusion as to which operation to do first, second, etc.

For example consider the expression:

$$3 * 2^3 - 2 + 6 \div 3$$

Now this looks simple enough, but it raises one important question: which operation do we perform first, second, third, etc.? To avoid this confusion and to ensure that we all get the same result when evaluating such expressions, a standard set of rules has been established. It is very critical that one understands this, since the computer is set up to use this protocol and regardless of what we might intend, it will calculate a result based upon this set of rules.

The order of operations states that we do the operation(s) inside of the **P**arentheses first. If we have nested parentheses (parentheses within parentheses), we do the innermost parentheses first and work outward. Next we do all **E**xponent operations. Any calculation of roots or radicals is done at the same level as exponents. After the exponents, we do all **M**ultiplication or **D**ivision simultaneously (whichever comes first) from left to right in the expression. Finally we do the **A**ddition or **S**ubtraction simultaneously from left to right. The order of operation can sometimes be remembered by the acronym **PEMDAS**.

Now, let us solve the problem above applying the order of operations:

$$3 * 2^3 - 2 + 6 \div 3$$

$3 * 2^3 - 2 + 6 \div 3$	raise 2 to the third power
$= 3 * 8 - 2 + 6 \div 3$	multiply 3 and 8
$= 24 - 2 + 6 \div 3$	divide 6 by 3
$= 24 - 2 + 2$	subtract 24 and 2
$= 22 + 2$	add 22 and 2
$= 24$	

Let us do some more examples showing how the order of operations is applied.

Example 1: Perform the indicated operations using the order of operations:
$5^2 - 3(2-6)$

Solution: Using the order of operations

$\quad 5^2 - 3(2-6)\quad$ perform the subtraction in the parenthesis 2-6

$= 5^2 - 3(-4) \quad$ square 5 or raise it to the second power

$= 25 - 3(-4) \quad$ multiply -3 by -4

$= 25 + 12 \quad\quad$ add 25 and 12

$= 37$

Example 2: Perform the indicated operations using the order of operations:
$(-5)(-2) - [3+(8-10)]$

Solution: Using the order of operations

$\quad (-5)(-2) - [3+(8-10)] \quad$ subtract 10 from 8 in the parenthesis

$= (-5)(-2) - [3+(-2)] \quad$ add 3 and -2

$= (-5)(-2) - (1) \quad\quad$ multiply -5 and -2

$= 10 - 1 \quad\quad\quad\quad$ subtract 1 from 10

$= 9$

Example 3: Perform the indicated operations using the order of operations:
$2 + 4[3+(2)(4)]$

Solution: Using the order of operations

$\quad 2 + 4[3+(2)(4)] \quad$ multiply 2 and 4

$= 2 + 4[3+8] \quad\quad$ add 3 and 8

$= 2 + 4(11) \quad\quad\quad$ multiply 4 and 11

$= 2 + 44 \quad\quad\quad\quad$ add 2 and 44

$= 46$

When we have fractions such as $\dfrac{3+9}{5-2}$ where there are operations that have to be performed either in the numerator or the denominator or both, the fraction line acts as a grouping symbol. In other words, we can write the fraction as $(3+9) \div (5-2)$ where we perform the operation in the numerator as well as the operation in the denominator, and then divide.

It is important to understand this because when you use a calculator or write a program, you must know where to insert a parenthesis even if you don't see one. The following is an example.

Example 4: Perform the indicated operations using the order of operations:
$$(-8+3)^2 - \frac{(5+10)}{7-2}$$

Solution: Using the order of operations

$\qquad (-8+3)^2 - \dfrac{(5+10)}{7-2}\qquad$ add -8 and 3

$\qquad =(-5)^2 - \dfrac{(5+10)}{7-2}\qquad$ add 5 and 10; subtract 2 from 7

$\qquad =(-5)^2 - \dfrac{15}{5}\qquad$ evaluate $(-5)^2$

$\qquad =25 - \dfrac{15}{5}\qquad$ divide 15 by 5

$\qquad =25-3\qquad$ subtract 3 from 25

$\qquad =22$

Example 5: Perform the indicated operations using the order of operations:
$$\left[4-3(-2)+5^2\right](-2)$$

Solution: Using the order of operations

$\qquad \left[4-3(-2)+5^2\right](-2)\qquad$ evaluate 5^2

$\qquad =[4-3(-2)+25](-2)\qquad$ multiply -3 and -2

$\qquad =[4+6+25](-2)\qquad$ add 4 and 6

$\qquad =[10+25](-2)\qquad$ add 10 and 25

$\qquad =(35)(-2)\qquad$ multiply 35 by -2

$\qquad =-70$

We discussed the absolute value and square root operators earlier in the chapter. The following example shows the importance of knowing when to use parenthesis in a problem.

Example 6: Perform the indicated operations using the order of operations:
$(-3)|6-8|+\sqrt{11-2}$

Solution: Using the order of operations

$\quad (-3)|6-8|+\sqrt{11-2}\quad$ evaluate the absolute value $|6\text{-}8|$
$=(-3)(2)+\sqrt{11-2}\quad$ subtract 11 and 2
$=(-3)(2)+\sqrt{9}\quad$ find the square root of 9
$=(-3)(2)+3\quad$ multiply -3 and 2
$=-6+3\quad$ add -6 and 3
$=-3$

How do we solve $5^2-3(2-6)$ using the computer?

Remembering that we use ^ as an exponent and the importance of parenthesis, we write the code as:

$\quad$ PRINT $\quad$ 5^2-3*(2-6)

A couple of comments are in order here. First, anytime there is multiplication, we need to explicitly use the * symbol. Failing to put the * symbol will result in error. Second, notice the parenthesis around $2-6$.

When writing the line for $(-3)|6-8|+\sqrt{11-2}$, make sure that a parenthesis surrounds the $6-8$ and $11-2$. The code for this problem should look like:

$\quad$ PRINT $\quad$ -3*ABS(6-8)+SQR(11-2) .

If the parenthesis is missing in ABS(6-8), then the absolute value of 6 is evaluated and then multiplied by -3, and then 8 is subtracted.

Example 7: Let a = 4 and b = 3. Write a program to evaluate the following expression using assignment statements and PRINT statement.
$(a+b)^2-2b$

Solution:
```
CLS
a = 4
b = 3
PRINT (a + b)^2 - 2*b
END
```

The output should be 43. Note that in the mathematical expression, the multiplication of 2 and b in 2b is understood. The program on the other hand must have the multiplication symbol * placed between the 2 and b.

EXERCISES 1.4

Simplify or evaluate:

1. $\sqrt{36}$
2. $\sqrt{121}$
3. $\sqrt{144}$
4. $\sqrt{100}$
5. $-\sqrt{49}$
6. $\sqrt{\dfrac{4}{9}}$
7. $\sqrt{\dfrac{441}{25}}$
8. $\sqrt{0.04}$
9. $-\sqrt{196}$
10. $\sqrt{\dfrac{225}{289}}$
11. $\sqrt[3]{64}$
12. $\sqrt[3]{-125}$

Use your calculator to find the square root of the following numbers. Round your answer to the nearest hundredths.

13. $\sqrt{20}$
14. $\sqrt{73}$
15. $-\sqrt{45}$
16. $\sqrt{\dfrac{4}{5}}$
17. $-\sqrt{\dfrac{5+9}{3+4}}$

What will be the output when of each of the following statements is executed?

18. `PRINT 100^(1/2)`
19. `PRINT 1728^(1/3)`
20. `PRINT 1024^(1/2)`
21. `PRINT SQR(20+16)`
22. `PRINT SQR(30)`
23. `PRINT 200^(1/3)`

State the property or properties used in each of the following exercises:

24. $(5+3)+7=5+(3+7)$

29. $19+37=37+19$

25. $(12+25)*3=12*3+25*3$

30. $(29*43)*4=29*(43*4)$

26. $42*66=66*42$

31. $21+(59+85)=(21+59)+85$

27. $38+94=94+38$

32. $8*\frac{1}{8}=1$

28. $64+0=64$

33. $2*(13+27+39)=2*13+2*27+2*39$

Solve each of the following percent problems.

34. What is 45% of 250?

35. 75 is what percent of 200?

36. 82 is 40% of what number?

37. If the sales tax rate is 8.75%, and if you purchased a computer for $960, how much sales tax do you have to pay?

38. Peter borrowed $200 from his friend and promises to pay an interest of 5% at the end of one year. How much will the interest be and how much money does Peter have to give his friend at the end of the year?

39. The selling price of a house went down from $400,000 to $375,000. What is the percent decrease of the selling price of the house?

Perform the indicated operations, using the order of operations.

40. $(5-3)^2 - 3^2$

41. $5 - 2\left[3 - (2-4)^2\right]$

42. $-3^2 + (7-5)^3 - 24 \div 6 * 3$

43. $\dfrac{16 \div 8 + 3 * 2}{\sqrt{64}} + 3(6-3)$

44. $32 \times 5 + 3^3 - (6^2 - 64 \div 4)$

45. $42 \div 2 \times 3 - 5 \times 3$

46. $3[4-(-3)] + 6\left[4^2 - (-7)\right]$

47. $8 - [2(3-4) - 7(1-3)]$

48. $5 - 2|1 - (4+2)|$

49. $4 - 3^2 + 6 \div 2(-5)^2$

50. $30 - (-4)^2 + 15 \div 5 \cdot (-2)$

51. $14 + 30 \div 5 \cdot 2$

52. $-3\sqrt{25} - (-2)(-5)$

53. $\dfrac{(-6-8)}{-7} - (-2-3)|-3-4|$

54. $\dfrac{6 - \left(\dfrac{8-2}{-3}\right) - (-3)^3}{-3-2}$

COMPUTER ACTIVITY

Student Name_____ Section_____ Date_____

Entering Complex Expressions in BASIC

Evaluate the expressions using your knowledge of order of operations and the calculator. Record your results in the spaces next to the activity.

Let A=2 B=5 C= −1 D=4 These values can be altered.

1. $Y = A + \dfrac{B}{C} + D$ (The solution is: $Y = 2 + \dfrac{5}{-1} + 4 = 2 - 5 + 4 = 1$)

2. $X = \dfrac{A+B}{C+D}$

3. $Z = B^2 - 2B + 3$

4. $W = \dfrac{4C}{B} - \dfrac{A}{5}$

5. $V = \dfrac{\left(\dfrac{A+5}{2} - \dfrac{2B-5}{4}\right)^2}{(A+B)^3}$

BASIC Programming

Using the arithmetic operators for exponentiation (^), multiplication(*), division(/), addition(+), subtraction(-), and square root (SQR( )), enter each of the expressions in the activities above in the computer using the assignment statement and evaluate each of the expressions with the given values of variables by running the program. Remember the use of parenthesis where it is necessary is very important. Compare your results with the above calculations. Re-write this program using **INPUT** statements so that you can enter other values and see whether the results you get from the program are the same as when you calculate using paper and pencil.

Computer Results:

1. y = A + B/C + D   *2.*

3.   *4.*   *5.*

COMPUTER ACTIVITY

Student Name_____ Section_____ Date_____

Entering Complex Expressions in BASIC

Evaluate the expressions using only the calculator or by hand. Record your results in the spaces next to the activity.

Let A=2   B=5   C= -1   D=4   These values can be altered.

1. $R = (A*B) + C$

2. $S = \dfrac{A+B+C}{D}$

3. $T = \dfrac{D+2B}{(A+C)^2}$

4. $M = \dfrac{A+B+C+D}{4}$

5. $N = \sqrt{\dfrac{(A+C)}{D}} - \dfrac{B^2}{4}$ (Note: square root function SQR(*expression*) is used in the first part of this expression.)

BASIC Programming

Using the arithmetic operators for exponentiation (^), multiplication(*), division(/), addition(+), subtraction(-), and square root (SQR( )), enter each of the expressions in the activities above in the computer using the assignment statement and evaluate each of the expressions with the given values of variables by running the program. Remember the use of parenthesis where it is necessary is very important. Compare your results with the above calculations. Re-write this program using **INPUT** statements so that you can enter other values and see whether the results you get from the program are the same as when you calculate using paper and pencil.

Computer Results:

1.   *2.*

3   *4.*   *5.*

Section 1.5: Decimal Representation of Real Numbers

Now that we understand the working with numbers Writing irrational numbers using their exact symbolic form, or the rational form of rational numbers is not always convenient or practical. When specific values are needed to solve a problem, or are to be entered into a computer, an alternative is to write what is called the decimal equivalent of the number. In this section we show how to write the decimal equivalent of a number along with its related floating-point and scientific form.

Objective A: Fixed-Point Numbers

The numbers we encounter on a daily basis are called base 10 numbers. They are built from the ten digits 0, 1, 2, 3, 4, 5, 6, 7, 8 and 9. Furthermore, the location of a digit in a number is significant. The location of the digit is related to a specific numerical value called the "place value" for that position in the number. For example the number 52.7 means we have 5 ten's, 2 one's and 7 tenths. The place values for this number are the ten's, one's and tenth's. In general, the place value increases by a factor of 10 (multiplied by) as we move towards the leftmost digit and decreases by a factor of ten (divided by) as we move towards the right.

Written in exponent form the place values are represented as follows:

...
10^3 = 1000   Thousands Place
10^2 = 100   Hundreds Place
10^1 = 10   Tens Place
10^0 = 1   Ones Place
.   Decimal Point
10^{-1} = 0.1   Tenths Place
10^{-2} = 0.01   Hundredths Place
10^{-3} = 0.001   Thousandths Place
...

We can also represent any decimal number in **Expanded Form**. For example:

1347.54   = 1*1000 + 3*100 + 4*10 + 7*1 + 5*0.1 + 4*0.01
           = $1*10^3 + 3*10^2 + 4*10^1 + 7*10^0 + 5*10^{-1} + 4*10^{-2}$

233.0151   = 2*100 + 3*10 + 3*1 + 0*0.1 + 1*0.01 + 5*0.001 + 1*0.0001
           = $2*10^2 + 3*10^1 + 3*10^0 + 0*10^{-1} + 1*10^{-2} + 5*10^{-3} + 1*10^{-4}$

Any rational number can be represented in decimal form. To find this representation we just perform long division dividing the numerator by the denominator.

Example 1: Express $\frac{5}{2}$ as a decimal number.

$$\frac{5}{2} \rightarrow 2\overline{)5} \rightarrow \begin{array}{r} 2.5 \\ 2\overline{)5.0} \\ \underline{-4} \\ 1.0 \\ \underline{-1.0} \\ 0 \end{array}$$

Example 2: Express $\frac{11}{3}$ as a decimal number.

$$\frac{11}{3} \rightarrow 3\overline{)11} \rightarrow \begin{array}{r} 3.6\overline{6} \\ 3\overline{)11.0} \\ \underline{-9} \\ 2.0 \\ \underline{-1.8} \\ .20 \\ \underline{-.18} \\ .02 \end{array}$$

Notice that in the example 1, the division terminates or ends. We call these type of decimal numbers terminating decimals. In example 2, the digit 6 is repeated. We call these types of decimals repeating decimals.

The numbers above are called fixed-point numbers, since the decimal point is placed at its natural location (between the one's and the tenths digits.) The decimal separates the whole number from the fractional part of the number and is always "fixed."

An irrational number can also be represented as a decimal number. The problem with irrational numbers, unlike the rational numbers, is that the representation is always approximate. This is the unique feature that separates the rational from the irrational numbers. Rational numbers either yield a terminating or repeating decimal representation, while irrationals always give a non-terminating, non-repeating decimal number. Consequently, it would take an infinite number of digits to exactly represent an irrational number, and this is not possible.

Examples:

Here are examples of fixed-point representation of irrational numbers:

$\pi = 3.141592...$, $\sqrt{2} = 1.4142...$, $e = 2.718281...$, $-\sqrt{5} = -2.2360679...$

We should again note that the decimal representation neither terminates nor does it start to generate a repeating pattern of digits.

Objective B: Floating-Point and Scientific Notation

In performing calculations, we frequently encounter numbers that are very large or that have a very small magnitude (nearly equal to zero). When we encounter very large or very small numbers, it is usually convenient to represent them in an alternative form called Scientific Notation.

In scientific notation, a number is written as $a * 10^n$, where $1 \leq a < 10$ and n is an exponent. The exponent n is positive if the number is greater than 10 and the exponent is negative if the number is less than 1.

If we moved the decimal point to the left we count the exponent as positive. This is because we would have to multiply our number by this many factors of ten, i.e. 10 raised to the positive exponent, to obtain the fixed-point representation of the number.

For example, consider the fixed-point number,

1,600.0

the scientific notation form would be

$1.6 * 10^3$.

The decimal point is placed between the 1 and 6 because we want exactly one digit to its left. Then 10^3 is used because 10^3 is equivalent to 10*10*10=1,000 and 1.6 times 1,000 is equal to 1,600.0, our original number. The two representations are equivalent forms of the same number.

If we have to move the decimal point to the right, the exponent will be negative. This is due to the fact that we are dividing by this many factors of 10, and we recall that exponentiation to a negative power is equivalent to division.

For example, consider the fixed-point number,

    0.00023   (23 hundred-thousandth's)

The scientific notation form would be $2.3 * 10^{-4}$

This is because 10^{-4} is equivalent to $\dfrac{1}{10^4} = \dfrac{1}{10*10*10*10} = \dfrac{1}{10,000}$ and multiplying 2.3 times this numbers yields our original number 0.00023. Again, the two representations are equivalent.

Example 1:   Write 1,230,000,000 in scientific notation.

Solution: $1,230,000,000 = 1.23*10^9$
(Since we moved the decimal point 9 places to the left.)

Example 2:   Write 0.000035465 in scientific notation.

Solution: $0.000035465 = 3.5465*10^{-5}$
(Since we moved the decimal point 5 places to the right.)

In working with computers we often encounter numbers in a slightly different form. Instead of putting the decimal place to the right of the first digit in the number(scientific notation), it is sometimes placed before the first digit, at the front of the number. These are called **floating-point numbers**.

Example 3:   Write 1,230,000 using floating-point notation.

Solution: $1,230,000 = 0.123*10^7$

Example 4:   Write 0.000035465 using floating-point notation.

Solution: $0.000035465 = 0.35465*10^{-4}$

Note: Oftentimes the zero to the left of the decimal point in the floating-point number is omitted. Either form is perfectly acceptable.

The 123 and the 35465 parts of the numbers above are called the mantissas and the 7 and the –4 are called exponents. Note that a positive exponent indicates that the decimal point of the mantissa is moved to the left; a negative exponent indicates it is moved to the right, just as it was for the scientific notation form.

Example 5: Write each number in scientific notation and floating-point notation:

Number	Scientific notation	Floating-point notation
245,000	$2.45*10^5$	$0.245*10^6$
0.00628	$6.28*10^{-3}$	$0.628*10^{-2}$
2,293,000	$2.293*10^6$	$0.2293*10^7$
0.0005527	$5.527*10^{-4}$	$0.5527*10^{-3}$

Example 6: The following numbers are expressed either in scientific notation or floating-point notation. Convert each of them to decimal (base 10) notation:

Number	Decimal notation
$1.88465*10^7$	18,846,500
$3.95*10^{-4}$	0.000395
$0.619*10^{-6}$	0.000000619
$0.342*10^5$	34,200
$4.9712*10^7$	49,712,000

EXERCISES 1.5

Write the decimal representation of each of the following numbers rounding each to the nearest hundredths.

1. $\dfrac{4}{15}$
2. $\dfrac{178}{42}$
3. $\sqrt{29}$
4. $-\sqrt{109}$

Write each number in a) scientific notation and b) floating-point notation:

5. 5,280,450,000
6. 0.0000967
7. 109,230,000,000
8. 55,000
9. 0.0000000534
10. 100,000,000,000

Convert each number expressed either in scientific notation or floating-point notation into decimal (base 10) notation:

11. $1.00007*10^7$
12. $0.499*10^5$
13. $5.536*10^{-5}$
14. $0.2005*10^{-3}$
15. $1.023*10^6$
16. $0.7744*10^8$

Section 1.6: Two Additional Common Operators: MOD and INT

MOD Operator

We developed our system of numbers based upon the concept of a number line that extends indefinitely in either direction. Not all systems satisfy this condition, however. There are many systems that do not give rise to indefinite values if left to continue to increase or decrease without end. Take for example a clock. As the clock continues to run, it repeats itself every 12 hours. Another example is the odometer on an automobile. After a fixed number of miles the odometer will return to zero and then start counting from the beginning again. Also, the 12 months in a year, or days in a week. Can you think of others?

These are examples of what we call modular systems. Instead of increasing without bound these systems cycle through a fixed set of numbers. Instead of a horizontal line with two arrows on either end, a modular system can be viewed as a circle. After it goes around the circle once, it begins to repeat itself. See the figure below.

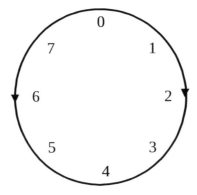

We will on occasion find it necessary to work with modular systems. Thus, we need to develop some tools that will enable us to analyze and operate in such systems. The most fundamental tool is the MOD operator. Sometimes when we are working with modular systems we are not concerned with how many times we may have gone around the circle, but only where we end up. For example, let's say you are paying on a loan and you have made 69 payments and the first payment made was in January. What is the last month you made a payment? We know there are 12 months in a year, thus we are not concerned with all the multiples of 12 in 69, but only the remainder, i.e.

69/12=5 Remainder 9

This result says that we went around the monthly clock 5 times (years) and the remainder term says that we made 9 additional monthly payments beyond that bringing us to the 9th

month or September. As our example illustrates, this can be a useful operation to have. This operation is called the modular operator, written as MOD, and is used to find the remainder term when dividing two whole numbers.

Thus, 69 MOD 12 = 9

Example 1: Evaluate 36 MOD 3.

  Solution: 36 MOD 3 = 0 since 36/3=12 Remainder 0

Example 2: Evaluate 15 MOD 7

  Solution: 15 MOD 7 = 1 since 15/7 = 2 Remainder 1

Example 3: Evaluate 6 MOD 8

  Solution: 6 MOD 8 = 6 since 6/8 = 0 Remainder 6

INT Operator

We introduce here another common operator used when writing programs. The **INT** operator is used on a given number to find the greatest integer less than or equal to that number. The mathematical symbol for this operator is $\lfloor x \rfloor$ where x is a real number. So, **INT**$(x) = \lfloor x \rfloor$ = greatest integer less than or equal to x. The greatest integer function will be discussed in detail in chapter 4.

Example 1: Evaluate INT(4.7)
  Solution: INT(4.7) = 4 since 4 is the greatest integer less than or equal to 4.7

Example 2: Evaluate INT(3)
  Solution: INT(3) = 3 since 3 is the greatest integer less than or equal to 3

Example 3: Evaluate INT(-2.3)
  Solution: INT(-2.3) = -3 since -3 is the greatest integer less than or equal to -2.3

EXERCISES 1.6

Evaluate each of the following:
1. 25 MOD 3
2. 67 MOD 8
3. 103 MOD 11
4. 58 MOD 7
5. 6 MOD 8
6. 39 MOD 6
7. 49 MOD 5
8. 100 MOD 5
9. 47 MOD 4
10. 28 MOD 3
11. 254 MOD 12
12. INT (4.5)
13. INT (-3.2)
14. INT(4.6 – 3.7)
15. INT(50/4)
16. INT(12.5*1.9)

COMPUTER ACTIVITY

Student Name_____ Section_____ Date_____

1. Complete the table by filling in the value of each variable after each line of program is executed. Fill the OUTPUT box with what you will see in your screen.

Program	A	B	C
INPUT " A = "; A			
INPUT " B = "; B			
C = INT(A/B)			
PRINT C			
C = (2*A + B) MOD B			
PRINT C			
A = 2*A			
PRINT A, B, C			
END			

Use A = 9 and B = 4

What should your screen look like when you RUN the program?

OUTPUT:

Section 1.7: Additional Programming Concepts

We close this chapter by writing **BASIC** programs that use the mathematical concepts we have developed. We will also increase our knowledge of programming by examining the various parts of a program and the data flow within a program.

The appendix lists the **BASIC** statements most commonly used in this text. A complete list of **BASIC** statements can be found on the computer when you access the **HELP** menu when you are in **BASIC**. You can always refer to the **HELP** screen for reference.

The following is a summary of some important facts when writing **BASIC** programs:

I. Arithmetic Operator symbols used when writing mathematical expressions in BASIC programs
   ^Exponentiation
   *Multiplication
   / Division
   +Addition
   − Subtraction

II. Some built-in BASIC operators
   ABS   Absolute value
   SQR   Square root
   MOD   Modulo
   INT   greatest integer less than or equal to

III. Know the order of operations
   1. Parenthesis
   2. Exponents (including roots and radicals)
   3. Multiplication or division from left to right
   4. Addition or subtraction from left to right

IV. Program development sequence
   1. Analyze the problem.
   2. Design a program to solve it.
   3. Write the code for the program.
   4. Test the program and debug if necessary.

The following program illustrates the difference between printing a list separated by semicolons and printing a list separated by commas:

OUTPUT:

```
CLS                         50 30 1500
L = 50                      160            50           30
W = 30
A = L*W
PRINT   L; W; A
P = 2*L + 2*W
PRINT   P, L, W
END
```

The *variables* in this program are L for length, W for width, P for perimeter, and A for area. Specifically, these *variables* are called *numeric variables* since each contains a number. We call these statements **assignment statements**. The number 50 is assigned to L (length) and 30 is assigned to W(width), while the product of the length and the width is assigned to A (area). The sum of twice the length and twice the width is assigned to P (perimeter).

Notice that the first PRINT statement's variables L, W, and A are separated by semi-colons while the second PRINT statement's variables are separated by commas. Using commas to separate variables will print the items in preset print areas usually 14 spaces per print field. Using semicolon to separate the variables will print the items close to each other. Refer to the PRINT statement in the **BASIC** statement in the appendix for more explanation of the difference between the use of commas and semi-colons.

There are two types of **variables**: numeric and character string. The comments below apply to **BASIC** language in general, they may be different for different versions of **BASIC** and for other programming languages.

- Variable names can be from 1 to 40 characters long and must begin with a letter of the alphabet.

- Variable names consist of letters, numbers, and periods (blanks are not allowed).

- Do not use reserve words such as PRINT, END, or OPEN.

- The assignment statement should be
  variable = numeric expression, or
  variable = number or
  variable = numeric variable expression.

- For character strings (usually called strings), the variable name must end with a **$**.

- A string variable can be declared as string at the beginning of the program so the $ sign does not have to be appended at the end of the variable name.

    An example of string variable assignment is:

    ```
    name$ = "Mary Smith"
    ```

    OR

    ```
    DIM name as STRING
    Name = "Mary Smith"
    ```

The *string variable* name$ is assigned a value Mary Smith, the name of a person. Notice that "Mary Smith" is enclosed in quotation marks. These quotation marks will not appear when the *string variable* "Mary Smith" is printed. The quotation marks are part of the syntax of assigning a string of characters to a variable. Quotation marks can also be used to enter dates where there are commas, such as "December 23, 1961".

More on INPUT and READ/DATA

Assignment statements are not the only way of assigning values to variables. There are at least two other methods. They are user supplied response INPUT statement or using READ/DATA statements. We have already discussed user supplied response INPUT statements in section 1.2. We'll give another example.

```
CLS
INPUT "Enter the price of the sweater"; price
total = price*1.0825
PRINT  "The price of the sweater is "; price;" and
        the total amount you pay including tax is ";total
END
```

Try the above program in your computer and compare your answer with your calculation.

OUTPUT:

```
                        Price of the sweater is
                                              ╱ Entered by user
┌─────────────────────────────────────────────────────────────────┐
│ Enter the price of the sweater? ____                            │
│ The price of the sweater is _____ and the total amount you pay including tax is _____ │
└─────────────────────────────────────────────────────────────────┘
        Price of the sweater                   Total amount paid
```

NOTE: If you want just the sentence

The price of the sweater is _____ and the total amount you pay including tax is _____

to show on the output screen, then insert **CLS** just before the **PRINT** statement.

READ/DATA statements: The variables in the **READ** statement are assigned values that are in the **DATA** statement. They are assigned in the order in which they occur. Numeric and character strings variables must be matched with appropriate data type. **READ** statements appear when a variable needs an assignment of data. **DATA** statements can appear anywhere in the program, but generally are grouped together just before the **END** statement.

Example 1:

```
CLS
READ   name$, grade1, grade2
PRINT  name$; grade1; grade2
READ   name$, grade1, grade2
PRINT  name$; grade1; grade2
DATA   "Mary Smith", 85, 73
DATA   "John Doe", 70, 92
```

OUTPUT:
Mary Smith 85 73
John Doe 70 92

Some different ways of executing **READ/DATA** statements:

Method 1	Method 2	Method 3
READ A,B,C	READ A	READ A
.	READ B	READ B
.	READ C	READ C
. or	. or	.
.	.	.
DATA 50	DATA 50	
DATA 150	DATA 150	DATA 50,150,250
DATA 250	DATA 250	

The three different sets of codes will each result in the same assignment of 50 to A, 150 to B and 250 to C.

RESTORE statement: Sometimes data will be read more than once; in this case, the statement **RESTORE** is used. Once this statement is executed, the program will return to the first value of the **DATA** statements the next **READ** statement is executed.

Example 2:

```
REM   Example of using RESTORE statement
   READ   num1,num2,num3
   RESTORE
READ num4,num5,num6,num7
PRINT   num1,num2,num3,num4,num5,num6,num7
DATA    5,10,15,20,25,30,35
```

OUTPUT:

5 10   15   5 10   15   20

Example 3:

```
REM   the output without using the RESTORE statement
READ   num1,num2,num3
READ num4,num5,num6,num7
PRINT   num1,num2,num3,num4,num5,num6,num7
DATA    5,10,15,20,25,30,35
```

OUTPUT:
  5   10   15   20   25   30   35

The following three programs illustrate three ways to write a program to calculate the salary given the hourly rate and number of hours worked. We can see how the next program is better written than the previous one.

```
REM      This program will calculate weekly salary
REM      Input the numbers of hours worked and the hourly
         rate
INPUT    hours,rate
LET      salary = hours * rate
PRINT    hours, rate, salary
END
```

Enter this program and run it.

What do you observe?_____

 What do you see on the output screen?_____

You most probably saw just a ? sign. It is really waiting for an input. But if the user running this program does not know what it is that he/she is being asked to input, then the program is not well-written.

Let us improve the code:

```
REM      This program will calculate weekly salary
REM      Input the numbers of hours worked and the
         hourly  rate
PRINT    "Enter the number of hours worked"
INPUT    hours
PRINT    "Enter the hourly rate"
INPUT    rate
salary = hours * rate
```

```
REM      To identify what is being printed,
         include descriptions
PRINT    "hours worked: ";hours, "hourly rate: ";rate,
         "weekly salary:";salary
END
```

Modify your program and run it.

Does it make more sense?_____

A different way of writing this code is:

```
REM      This program will calculate weekly salary
REM      Input the numbers of hours worked and
         the hourly rate
INPUT    "Enter the number of hours worked";hours
INPUT    "Enter the hourly rate";rate
salary = hours * rate
PRINT    "hours worked: ";hours, "hourly rate: ";rate,
         "weekly salary:";salary
END
```

The program above simply merged the PRINT and INPUT statements into a single statement. There is no dollar sign in the salary nor has it been rounded to two decimal places, currency formatting is discussed in the appendix under PRINT USING.

COMPUTER ACTIVITY

1. Complete the table by filling in the value of each variable after each line of program is executed. Fill the OUTPUT box with what you will see in your screen.

Program	A	B	C
INPUT "A = "; A	9		
INPUT "B = "; B	9	4	
C = INT(A/B)	9	4	2
PRINT C	9	4	2
C = (2*A + B) MOD B	9	4	2
PRINT C	9	4	2
A = 2*A	18	4	2
PRINT A,B,C	18	4	2
END			

Use A = 9 and B = 4

OUTPUT:

```
2
2
18   4   2
```

2. What will be the output of the following program?

```
READ      a, b$
READ      c$, d
PRINT     a,b$,c$,d
RESTORE
READ      d
PRINT     a,d
DATA      8, Margaret, Kerwin, 21
END
```

OUTPUT:

```
8   Margaret   Kerwin   21
8   8
```

3. Re-write the following program using Method 2 in the READ/DATA section.

   ```
   CLS
   READ  name$, grade1, grade2
   PRINT name$; grade1; grade2
   READ  name$, grade1, grade2
   PRINT name$; grade1; grade2
   DATA  "Mary Smith", 85, 73
   DATA  "John Doe", 70, 92
   ```

4. What will be the output of the following program?

   ```
   A = 3
   B = 5
   A = INT(B/A)
   PRINT A
   PRINT A + B
   ```

   OUTPUT:

   ```
   ┌─────────────────────────────────────────┐
   │                                         │
   │                                         │
   └─────────────────────────────────────────┘
   ```

5. Complete the table by filling in the value of each variable after each line of program is executed. Fill the OUTPUT box with what you will see in your screen.

Program	A	B
A = 5		
B = 6*A/2 - A		
PRINT B		
B = 6*A/(2 - A)		
PRINT B		
A = -B		
PRINT A		
PRINT -A		
END		

   What should your screen look like when you RUN the program?

OUTPUT:

6. Complete the table by filling in the value of each variable after each line of program is executed. Fill the OUTPUT box with what you will see in your screen.

Program	X
X = 1	1
X = X + 1	2
X = X * (X + 1)	6
PRINT X	6
X = X + X MOD 4	8
PRINT X	8
X = (X + X) MOD 4	0
PRINT X	0
END	0

What should your screen look like when you RUN the program?

OUTPUT:

6
8
0

Chapter 2

Numbers and Computers

In Chapter 1 we focused on developing a good "mathematical number sense." We worked with the familiar decimal (base 10) number system and reviewed common arithmetic operations on real numbers. Computers, on the other hand, like many electronic devices are designed around a totally different system of numbers called the binary number system. As we shall see, binary numbers are a natural consequence of electronic devices.

The purpose of this chapter is to provide a good "computer number sense." To accomplish this we begin by discussing how numbers are represented in a computer. This will help us to understand some of the important limitations a computer has. We presented a little of this when we discussed floating–point numbers in Chapter 1. Here we shall take it a step further and highlight some important consequences related to limitations when representing and working with numbers in a computer.

In this chapter, we will provide an introduction to the binary number system. We will explain the relation between binary, decimal and other base systems and show the basics of arithmetic operations involving them. For those working with computers, a thorough understanding of binary and other related number systems is indispensable. As we shall see, computers represent numbers and carry out calculations in a very different way internally.

After completing this chapter you should have obtained a good "computer number sense." More specifically, you should be able to:

- understand some basic computer terminology and convert between the various units of measurement,
- write whole decimals as either a binary, octal, or a hexadecimal number and vice–versa,
- write a fractional decimal number as a binary or hexadecimal and a fractional binary or hexadecimal as a decimal,
- add and subtract unsigned binary and hexadecimal numbers,
- use the two's complement method to write a signed binary number, and
- add or subtract signed binary numbers.

Section 2.1: Computer Terminology and Units of Measurements

The field of computers has produced a unique set of terminologies and abbreviations. Some of the terminology should already be familiar to you and has evolved into and become a part of our common language. It is almost impossible not to see an advertisement mentioning the speed of a processor or a personal organizer or Palm Pilot (MHz), the picture quality of a digital camera (mega pixels), or the capacity of a hard drive, compact disk, compact flash card, or DVD (MB). These terms have to do with memory size, storage capacity, and processor speed. In this section we focus on some of the important terms and units used when discussing computers.

Memory/Storage
Bit: A binary digit (either 1 or 0)
Byte: 8 bits
K: *   1024 when talking about Random Access Memory (RAM), otherwise 1000
KB: *   Kilobyte, 10^3 bytes=1000 bytes, or 2^{10}=1024 bytes
MB: *   Megabyte or "meg", equals 10^6 bytes = 1,000,000 bytes, or 2^{20}= 1,048,576 bytes, or $1,000*2^{10}$=1,024,000 bytes
GB: *   Gigabyte or "gig", equals 10^9 bytes = 1,000,000,000 bytes, or 2^{30}= 1,073,741,824 bytes, or $1,000,000 \times 2^{10}$=1,024,000,000 bytes
KiB: Kibibyte (kilo–binary), 2^{10} = 1024 bytes
MiB: Mebibyte (mega–binary), 2^{20}= 1,048,576 bytes
GiB: Gibibyte (giga–binary), 2^{30}= 1,073,741,824 bytes
RAM: Random Access Memory
ROM: Read Only Memory
BPI: Bytes per inch (used when discussing how data is stored on a magnetic tape)

Measurements of Time
Millisecond: One one–thousandth (10^{-3}) of a second (msec)
Microsecond: One one–millionth (10^{-6}) of a second (μsec)
Nanosecond: One one–billionth (10^{-9}) of a second (nsec)
Picosecond: One one–trillionth (10^{-12}) of a second (psec)

Processor Speeds
bps: bits per second
KOPS: Kilo Ops, 1,000 operations per second
MHz: Mega–Hertz (one–million, 1,000,000 cycles per second)
GHz: Giga–Hertz (one–billion, 1,000,000,000 cycles per second)

MTBF: Mean time between failures (hours/failure)

*Care must be taken whenever you see these designations. See the discussion in Objective A below for further clarification.

Many of the designations use a system of units based upon the international system of units (SI). This system of units uses modern metric designations for very large and very small numbers. The numbers along with their prefixes and symbols are presented in the table below. We shall use this table as a reference throughout this chapter.

SI Prefixes

Multiplication Factor		Prefix	Symbol
1 000 000 000 000 000 000 000 000	$= 10^{24}$	yotta	Y
1 000 000 000 000 000 000 000	$= 10^{21}$	zeta	Z
1 000 000 000 000 000 000	$= 10^{18}$	exa	E
1 000 000 000 000 000	$= 10^{15}$	peta	P
1 000 000 000 000	$= 10^{12}$	tera	T
1 000 000 000	$= 10^{9}$	giga	G
1 000 000	$= 10^{6}$	mega	M
1 000	$= 10^{3}$	kilo	k
100	$= 10^{2}$	hecto	h
10	$= 10^{1}$	deka	da
0.1	$= 10^{-1}$	deci	d
0.01	$= 10^{-2}$	centi	c
0.001	$= 10^{-3}$	milli	m
0.000 001	$= 10^{-6}$	micro	µ
0.000 000 001	$= 10^{-9}$	nano	n
0.000 000 000 001	$= 10^{-12}$	pico	p
0.000 000 000 000 001	$= 10^{-15}$	femto	f
0.000 000 000 000 000 001	$= 10^{-18}$	atto	a
0.000 000 000 000 000 000 001	$= 10^{-21}$	zepto	z
0.000 000 000 000 000 000 000 001	$= 10^{-24}$	yocto	y

Note: We will typically use values ranging from nano (10^{-9}) to giga (10^9), but as computers get faster and memory gets larger, we will undoubtedly encounter both smaller and larger units of measurements. Also note the use of the Greek letter µ (pronounced mu) for the micron. This is the only letter in the chart not from the English alphabet.

Objective A: Bits and Bytes

A bit is an electrical element within the computer, which can be either "on" or "off. " In electronic terms, it is a semi–conductor, which is able to conduct a tiny amount of electricity when it is "on" but cannot do so when it is "off. " When it is "on" it can be regarded as having the value *one* and in computer language the bit is then said to be "set." When it is "off" it can be regarded as having the value *zero* and in computer language the bit is "clear." Bits can only be set or clear and cannot have any other state. Therefore a bit can only have a value of 1 or 0. This is called a binary number.

Two or more bits can be joined together to create larger numbers. When bits are joined together, the bit on the right is the least significant and if it is "set" it represents the value one. The next bit to the left is more significant by a factor of two. When that bit is "set" it represents the value of two. Suppose you have a number formed with two bits. The only possible arrangements of two bits can be 00 or 01 or 10 or 11. The equivalent numbers in decimal form are 0, 1, 2, and 3 respectively (we will explore this in greater detail later in the chapter, but for now try and follow as best as you can).

A "byte" is 8 bits and a "word," depending upon the computer, is 16 bits (2 bytes), 32 bits (4 bytes) or 64 bits (8 bytes). Bytes and words are the basic chunks of data used in computer processing. Typical computer programs and applications may require many thousands or even millions of bytes of storage. It would be cumbersome to always have to write the exact decimal amount, thus, a system of abbreviations for large amounts of data was created. It turns out that there are two opposing systems for measuring the memory/storage size of a computer.

A word of caution and explanation is in order here.

The first is the familiar metric system and the second is its binary equivalent. The two systems are frequently used interchangeably or are mixed together creating hybrid systems. This can cause a great deal of confusion. In this section we will try to clarify the situation and also let you know both the correct and the generally accepted (but not necessarily correct) memory/storage designations.

The memory and storage space in a computer is built around the binary number system, which we will study in section 2.2. Early on it was decided to use the metric designation to represent large chunks of memory. Specifically, the metric prefix of kilo was chosen. In the metric world, a kilo stands for $1,000=10^3$. The problem is that in the binary world we only have multiples of two and not ten, like we have in the decimal or metric system. Thus, it was decided to merge both systems and choose the multiple of two that was closest to the value of 1000 in the metric system and call this unit a kilobyte. That number is $2^{10}=1024$.

As the memory size of computers grew, the metric designations of mega and giga were adopted in the same way. A megabyte was defined as the multiple of two that is closest to 1,000,000, which is $2^{20}=1,048,576$ bytes. A gigabyte was defined to be $2^{30}=1,073,741,824$ bytes. This, however, was in direct conflict with the metric system where a kilobyte is really 1,000 (thousand) bytes, a megabyte is 1,000,000 (million) bytes and a gigabyte is 1,000,000,000 (billion) bytes. The confusion, however, did not stop here. Some groups decided to define the megabyte and the gigabyte as simply 1,000x1024 (or a thousand kilobytes), and 1,000,000x1024 (or a million kilobytes), respectively. To date this still causes a great deal of confusion.

In December 1998 the International Electrotechnical Commission (IEC), the leading international organization for worldwide standardization in electrotechnology, approved a new designation for these binary numbers. This system is no longer in conflict with the metric system. They adopted a kilo–binary to be $2^{10}=1024$, and called this a kibibyte. The mega–binary was defined to be $2^{20}=1,048,576$ bytes, and called it a mebibyte. The giga–binary is simply $2^{30}=1,073,741,824$ bytes or gibibyte. They simply chose to take the first two letters in the metric designation along with the first two letters of the word "binary", i.e. Ki–Bi, Me–Bi and Gi–Bi. Although this was decided in the late 1990's it is, however, not widely used today. Just as it is hard for the metric system to be adopted, it is equally as hard for some people to change to this new designation system. To date this is still a problem and is reflected in the following chart:

Value of a "Megabyte"	Exponential Equivalent	Use Today
1,048,576	2^{20}	Computer scientists. Used in computer programs as a measure of memory size. Computer memory (RAM) chip manufacturers and CDs
1,000,000	10^6	Makers of hard disks and in some other components such as network hardware.
1,024,000	2^{10}x1,000	Used for floppy disks where 1.44MB has a capacity of 1,474,560 bytes instead of ≈ 1,509,949 bytes or 1,440,000 bytes

In this text we will always try to make it clear which system of units we are using. However, be forewarned, when you go out into the field, "what–you–see" may not be "what–you–get." If it is important, always find out directly from the vendor or manufacturer which designation they are using in their product specifications.

To a computer scientist 1 KB (kilobyte) of memory or storage space corresponds to the number two (binary unit), raised to a power that gives a number that is closest to 1 thousand (1,000). In this case it is $2^{10}=1024$. This is a basic building block of memory on a computer for computer scientists and for the makers of memory chips (not for the makers of hard drives, though). Similarly 1 MB (megabyte) corresponds to $2^{20}= 1,048,576$. Thus, a computer that has 3MB of RAM actually has 3*1,048,576= 3,145,728 bytes of RAM.

The relationship between the common memory sizes is illustrated in the table on the next page. To use this table, simply find the unit you are converting from in the leftmost column of the table, then locate the units you are converting to in the top row. The box where the column and row intersect is the conversion factor.

We should point out that the table has two different conversion factors for a MB and a GB. These are denoted MB_1, MB_2, GB_1 and GB_2. This is due to the mixed conversion system, where instead of a MB being equal to either 10^6 or 2^{20}, a MB is sometimes considered to be to equal to 2^{10}x1,000. In this book we will write either MB or MB_1 to identify 10^6 bytes and MB_2 to correspond to 2^{10}x1,000=1,024,000 bytes. A MiB will always mean 2^{20} bytes. A similar approach will be adopted when referring to GB, GB_1 and GB_2.

Memory Conversion Chart

	Bytes	KB	KiB	MB_1	MB_2	MiB	GB_1	GB_2	GiB
Bytes	1	0.001	$9.77*10^{-4}$	$1.0*10^{-6}$	$9.77*10^{-7}$	$9.54*10^{-7}$	$1.0*10^{-9}$	$9.77*10^{-10}$	$9.31*10^{-10}$
KB	1,000	1	$9.77*10^{-1}$	$1.0*10^{-3}$	$9.77*10^{-4}$	$9.54*10^{-4}$	$1.0*10^{-6}$	$9.77*10^{-7}$	$9.31*10^{-7}$
KiB	1,024	1.024	1	$1.024*10^{-3}$	$1.0*10^{-3}$	$9.77*10^{-4}$	$1.024*10^{-6}$	$1.0*10^{-6}$	$9.54*10^{-4}$
MB_1	1,000,000	1,000	977	1	$9.77*10^{-1}$	0.954	0.001	$9.77*10^{-4}$	$9.31*10^{-4}$
MB_2	1,024,000	1,024	1,000	1.024	1	0.977	$1.024*10^{-3}$	0.001	$9.54*10^{-4}$
MiB	1,048,576	1,049	1,024	1.049	1.024	1	$1.024*10^{-3}$	$1.024*10^{-3}$	$9.77*10^{-4}$
GB_1	1,000,000,000	1,000,000	976,563	1,000	977	954	1	$1.024*10^{-3}$	0.931
GB_2	1,024,000,000	1,024,000	1,000,000	1,024	1,000	977	1.024	1	0.954
GiB	1,073,741,824	1,073,742	1,048,576	1,074	1,024	1,000	1.074	1.024	1

Example 1: How many bytes are in 16 MB of memory (RAM)?

Solution: **Method 1**: Conversion table. In this approach we look for the appropriate conversion factor in the table and then multiply to obtain the answer.
We are converting from MB (MiB from the table, since the memory is RAM), to bytes. Thus, we use the conversion factor of 1,048,576, or

$$16 \text{ MB} = 16*1,048,576 \text{ bytes} = 16,777,216 \text{ bytes}$$

Method 2: Unit cancellation. In this approach we use the values of the various units and cancel out units until we obtain the final units we are looking for.

$$16 \text{ MB} = 16 \text{ \cancel{MiB}} * \frac{1,048,576 \text{ bytes}}{\cancel{MiB}} = 16,777.216 \text{ bytes}$$

Example 2: A computer has 67,108,864 bytes of RAM (Random Access Memory). How many MiB ($1,048,576 = 2^{20}$ bytes) does it have?

Solution: Without calculations, we could estimate that our answer should be near 67MiB since 67,108,864 is about 67 million and 1MiB is about one million.

Method 1: Conversion table
We are converting from bytes to MiB. Thus, we use the conversion factor of 9.54×10^{-7}, or

67,108,864 bytes = 67,108,864 *9.54×10^{-7} MiB = 64.02MiB. The conversion chart is only accurate to three significant figures, thus we round it to 64MiB.

Method 2: Unit cancellation
To compute the exact answer, we know that for RAM we use
1 MiB = 1,048,576 bytes. Divide 67,108,864 by 1,048,576 to obtain 64MiB.

$$67,108,864 \text{ bytes} = 67,108,864 \text{ bytes} * \frac{1 \text{ MiB bytes}}{1,048,576 \text{ bytes}} = 64 \text{MiB}$$

Example 3: A 40GB computer has how many bytes of hard drive storage space?

Solution: Since this is storage space we use GB_1=1,000,000,000 bytes.

Method 1: Conversion table

We are converting from GB_1 to bytes. Thus, we use the conversion factor of 1,000,000,000, or

40*1,000,000,000 bytes=40,000,000,000 bytes
Method 2: Unit cancellation

$$40GB = 40GB * \frac{1,000,000,000 \text{ bytes}}{GB} = 32,000,000 \text{ bytes}$$

Objective B: Measurements of Time

A modern computer can perform millions of operations per second. Thus, the time required to perform a small number of calculations can be a very small fraction of a second. To account for this, you will frequently see units of time in quantities other than seconds.

Some of the frequently used increments of time are the:
pico–second   psec   one–trillionth of a second   0.000000000001 seconds
nano–second   nsec   one–billionth of a second   0.000000001 seconds
micro–second   μsec   one–millionth of a second   0.000001 seconds
milli–second   msec   one–thousandth of a second   0.001 seconds

In the table below we list these units of time, along with their conversion factors.

	psec	nsec	μsec	msec	sec
psec	1	0.001	0.000001	0.000000001	0.000000000001
nsec	1,000	1	0.001	0.000001	0.000000001
μsec	1,000,000	1,000	1	0.001	0.001001
msec	1,000,000,000	1,000,000	1,000	1	0.001
sec	1,000,000,000,000	1,000,000,000	1,000,000	1,000	1

To use this table simply find the unit you are converting from in the leftmost column. Then find the unit you are converting to along the top row. The conversion factor is the number where the column and row intersect.

Example 1: How many μsec's are in 34 psec?

Solution: Since, we are converting **from psec**, we look at the psec row in the chart. We are converting **to μsec**, thus we look for the μsec column. This is the 1st

row and the 3rd column in the chart above. The value of the conversion factor is 0.000001, so we multiply by this amount. The answer is:

$$34*0.000001\mu sec=0.000034\mu sec=3.4\times 10^{-5}\mu sec$$

Example 2: How many nsec's are in 125 msec?

Solution: Since we are converting **from msec**, we look at the msec row in the chart. We are converting **to nsec**, thus we look for the nsec column. This is the 4th row and the 2nd column in the chart above. The value of the conversion factor is 1,000,000 (there are 1 million nano–seconds in a mili–second), so we multiply by this amount. The answer is:

$$125*1,000,000 msec=125,000,000 nsec$$

Example 3: Suppose a subroutine, a portion of a program that may be used several times, takes 0.0947 seconds (sec) to execute. How many microseconds (μsec) is it equivalent to?

Solution: Since we are converting **from sec**, we look at the sec row in the chart. We are converting **to μsec**, thus we look for the μsec column. This is the 5th row and the 3rd column in the chart above. The value of the conversion factor is 1,000,000 (there are 1 million micro–seconds in a second), so we multiply by this amount. The answer is:

$$0.0947*1,000,000\ \mu sec=94,700\mu sec$$

Objective C: Processor Speed

Another important quantity relates to the speed of the central processor of the computer. The processor is the heart of the computer and determines how many calculations the computer (processor) can perform in one second. A typical processor can perform an extremely large number of calculations every second. The metric system of units and abbreviations is used to represent these large numbers.

Currently the basic units of measuring processor speed are mega–Hertz (MHz) and giga–Hertz (GHz). The word (name) Hertz has a special meaning. It is used instead of saying cycles per second. It is in honor of a German physicist, Heinrich Rudolf Hertz, who in 1884 proved that electricity can be transmitted by electromagnetic waves which led to the development of wireless telegraph and the radio.

A 1 GHz machine can perform 1,000,000,000 (1 billion) cycles per second.

Example 1: A 3.6 GHz computer must execute code that requires 1,276,000,000 cycles. How many milli–seconds does it take to complete?

Solution:
$$\frac{1,276,000,000 \text{ cycles}}{3.6 \text{ GHz}} = \frac{1,276,000,000 \text{ cycles}}{3,600,000,000 \frac{\text{cycles}}{\text{second}}}$$

$$\approx 0.3544 \text{ seconds}$$

we now convert from seconds to msec as above to obtain: 354.4 msec.

Example 2: You need to run a program that requires approximately 3,564,000,000 cycles. The program must run in a half of a second. How fast does the computer processor have to be?

Solution:
$$\frac{3,564,000,000 \text{ cycles}}{0.5 \text{ seconds}} = 7,128,000,000 \frac{\text{cycles}}{\text{second}} = 7.128 \text{ GHz}$$

Is this possible using today's desktop computers?

Objective D: File Transfer Rate

The file transfer rate is a measure of how quickly a device can transfer data from one storage location to another.

Example 1: If we are given a tape drive and we know that it can move tape at a rate of 200 inches per second, and that the tape density is 2000 BPI (bytes per inch), then at what rate can the drive transfer or receive information?

Solution:
$$2000 \frac{\text{bytes}}{\text{in.}} * 200 \frac{\text{in.}}{\text{sec.}} = 400,000 \frac{\text{bytes}}{\text{sec.}}$$

Thus, the transfer rate of this device is 400 KB per second.

Example 2: A CD-ROM can transfer data at the rate of 7,800 KB/second (use 1KB=1,000 bytes). It has to transfer 350MB of data. How long will it take?

Solution:

$$\frac{350\text{MB}}{7,800\frac{\text{KB}}{\text{sec.}}} = \frac{350,000,000 \text{ bytes}}{7,800,000 \frac{\text{bytes}}{\text{sec.}}} \approx 44.87 \text{ seconds}$$

Thus, the total transfer time is 44.87 seconds.

Objective E: Disk Storage Capacity

The maximum amount of storage space on a storage media is called the disk capacity.

Example 1: A magnetic hard disk pack contains 10 hard disks. Each side of a disk is used for storage and each disk's surface is divided into 200 concentric circular tracks and eight pie–shaped sectors. Each track–sector can hold 500bytes of information (see figure below.) What is the total storage capacity of this disk pack?

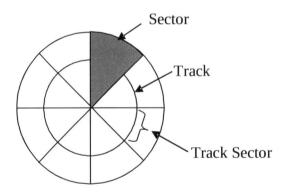

Solution: $10 \text{ Disks} * \frac{2 \text{ surfaces}}{\text{disk}} * \frac{200 \text{ tracks}}{\text{surface}} * \frac{8 \text{ track-sectors}}{\text{track}} * \frac{500 \text{ bytes}}{\text{track-sector}}$

$= 16,000,000 \frac{\text{bytes}}{\text{disk}}$

Thus, this pack can hold 16 MB of data.

Example 2: CDs store data in a spiral pattern on the disk, starting from the center and moving outward (see figure below). This creates one long continuous track to store data, that is approximately 5 millionths of a meter thick and 5 kilometers (3.5 miles) long. If the number of spirals (number of complete rotations) is doubled, then the length of the spiral roughly doubles the storage capacity of the CD. A CD that normally holds 750MB was redesigned to have twice as many spirals. How much data can the new disk hold?

Solution: Since we are doubling the number of spirals the length goes up by a factor of 2. This means that the storage capacity also increases by a factor of 2. Thus, we have

2*750 MB=1,500 MB=1.5 GB of storage capacity.

Objective F: Mean Time Between Failure

Another frequently used quantity is the mean time between failures (MTBF). When this term is applied to computers, this is the average time between failures on a computer. We compute the MTBF by simply dividing the total number of operating hours by the total number of failures that occurred during that time.

$$\text{MTBF} = \frac{\text{Total hours of operation}}{\text{Total number of failures}}$$

Example: A computer server has been operating for 7 days with a total of 12 failures. What is the MTBF?

Solution: 7 days is equal to, 7 days*24 hours per day=168 hours

$$\text{MTBF} = \frac{\text{Total hours of operation}}{\text{Total number of failures}} = \frac{168 \text{ hours}}{12 \text{ failures}} = 14 \frac{\text{hours}}{\text{failure}}$$

This means that on average there will be 1 failure every 14 hours.

EXERCISES 2.1

1. How many bytes are in 64K memory (RAM)? Express your answer as a power of 2.

2. How many bytes are in a 256K memory? Express your answer as a power of 2.

3. How many bytes and gigabytes (GB) are in 56 megabytes (MB) of memory?

4. How many bytes and gigabytes (GB) are in 1,024 megabytes (MB) of storage capacity?

5. A computer has 4,718,592 bytes of RAM, how many MiB and KiB does it have?

6. A computer has 11,534,336 bytes of RAM, how many MiB and KiB does it have?

7. A computer has a 33GB hard drive, how many bytes, KB and MB, KiB, MiB does it have?

8. A computer has a 80GB hard drive, how many bytes, KB and MB, KiB, MiB does it have?

Express the times in picoseconds (psec), nanoseconds (nsec), and microseconds (μsec).

9. 0.00045 sec
10. 892 psec
11. 3×10^5 μsec
12. 14.75 nsec
13. 0.000024 sec
14. 3,121 nsec
15. 4.6×10^6 μsec
16. 124.38 psec

17. A 3.2 GHz computer must execute code that requires it to perform 2^{30} operations. How many milliseconds (msec) does it take to compute?

18. A 2.5 GHz computer must execute code that requires it to perform 2^{28} operations. How many milliseconds (msec) does it take to compute?

19. A given tape drive can move tape at a rate of 100 inches per second. The tape density is 3600 BPI (bytes per inch). What is the rate the drive can transfer or receive information at?

20. A given tape drive can move tape at a rate of 750 inches per second. The tape density is 4800 BPI (bytes per inch). What is the rate the drive can transfer or receive information at?

21. A CD-ROM can transfer data at the rate of 9,600 KBps (kilo–bytes per second). It has to transfer 800 MB of data. How long will it take?

22. A CD-ROM can transfer data at the rate of 6,800 KBps (kilo–bytes per second). It has to transfer 750 MB of data. How long will it take?

23. A hard drive can transfer data at the rate of 48 MBps (mega–bytes per second). It has to transfer 900 MB of data. How long will it take?

24. A hard drive can transfer data at the rate of 74 MBps (mega–bytes per second). It has to transfer 3.4 GB of data. How long will it take?

25. A magnetic disk pack that contains 20 hard disks. Each side of the disk is used for storage and each disk's surface is divided into 300 concentric circular tracks and eight pie–shaped sectors. Each sector can hold 1 KiB of information. What is the total storage capacity of this disk pack?

26. A magnetic disk pack that contains 15 hard disks. Each side of the disk is used for storage and each disk's surface is divided into 150 concentric circular tracks and eight pie–shaped sectors. Each sector can hold 1.5 KiB of information. What is the total storage capacity of this disk pack?

27. A CD can currently store 750 MB of data. If the number of spirals (data storage track) is quadrupled, approximately how much data can the new disk hold?

28. A CD can currently store 1 GB of data. If the number of spirals (data storage track) is increased by a factor of 8, approximately how much data can the new disk hold?

29. A computer server has been operating for 2 weeks with a total of 15 failures. What is the MTBF?

COMPUTER ACTIVITY

30. A computer server has been operating for 30 days with a total of 62 failures. What is the MTBF?

1. Write a computer program that will input a memory size in units of bytes and output the number of KB, KiB, MB, MiB, GB and GiB.

2. Write a computer program that will input a memory size in MB units and output the number of bytes, KB, KiB, MiB, GB and GiB.

3. Write a computer program that will input a time in seconds and output the equivalent number of msec, μsec, nsec and psec.

4. Write a program that will input a length of time and the number of failures that occurred during that time and output the MTBF in hours/failure.

5. Write a program that will input the processor speed of a computer in addition to the number of operations the computer must perform and then output the time it takes the computer to perform these operations in msec.

6. Write a program that will input how fast a device can store and retrieve data in Bps (bytes per second) as well as a total number of bytes to be transferred in MB and output the total time this would take in msec.

Section 2.2: Working With Numbers in a Computer

Throughout chapter 1 we talked about numbers, their representations, and operations on them. We also discussed their relationship with a computer. In this section we begin to discuss how numbers are represented, stored, and processed in a computer. We also take a closer look at some of the problems that may arise.

Objective A: Representing Real Numbers in a Computer

Computers are very powerful and quite versatile, however, they do have limitations. The first important limitation is that every computer has a fixed and finite amount of memory and storage. Thus, there are limitations on the size and type of number that can be stored in a computer. It is not possible to store every real number in a computer. You cannot store any irrational number "exactly" in a computer. A computer has only a finite amount of space to store a number and an irrational is equivalent to a non–terminating, non–repeating decimal. We would be required to store an infinite number of digits. The obvious question then is, how this is done?

The first thing we need to be aware of is that the storage space allocated for a number can vary from computer to computer. The details of this will be explained more fully in your computer courses. For our purposes, it will be sufficient to look at the impact it has on the numbers directly. Although computers work in binary numbers, we shall illustrate this in the more familiar base 10 number system. The concepts are the same, but it will be easier to follow in decimal than binary.

Let's consider a very limited computer that will only allow us to store our number in a register that is 8 digits (or characters) long as shown below.

What if we wanted to store the number 97,428,395 on the computer? As we can see by the figure below, this does not pose a problem, since we can store the entire number in our 8 digit register.

What about the number 0.000001483? Here we'd have a problem. Counting the zeros, we have nine digits as well as the decimal point to account for. Obviously we cannot store this number in its present fixed–point form. Recall from Section 1.7 how we represent a number in floating–point form, i.e.

$$0.000001483 = .1483 \times 10^{-5}$$

Now if we adopt a convention that the first 5 register locations from the left are reserved for the sign of our number and its most significant digits to the right of the decimal point (the mantissa), and that the last three register locations are reserved for the exponent and its sign, we can store the number as:

+	1	4	8	3	-	0	6

Thus, we can accurately represent this number in the computer. However, this convention does not allow us to represent the number 97,428,395 from example 1 in the computer.

What about the number –3,453,000,000,000? Again we can write this as a floating–point number:

$$-3,453,000,000,000 = -.3453 \times 10^{13}$$

Then we can fit this in out 8–bit register as:

-	3	4	5	3	+	1	3

Examples: Given a computer with an 8-digit register, show how the following numbers would be stored:

1. 8,320,000,000,000
2. –0.000 000 975
3. –35,390
4. 0.000 774

Solutions:
1. Rewrite 8,320,000,000,000 in floating–point form as $.832 \times 10^{13}$ the mantissa is 832, the exponent is 13 and the sign of the number is positive. Therefore we can write the number in the 8-digit register as:

+	8	3	2	0	+	1	3

2. Rewrite –0.000 000 975 in floating–point form as $-.975\times10^{-6}$
   The mantissa is 975, the exponent is –6 and the sign of the number is negative. Therefore we can write the number in the 8-digit register as:

-	9	7	5	0	-	0	6

3. Rewrite –35,390 in floating–point form as $-.3539\times10^{5}$
   The mantissa is 3539, the exponent is 5 and the sign of the number is negative. Therefore we can write the number in the 8-digit register as:

-	3	5	3	9	+	0	5

4. Rewrite 0.000 774 in floating–point form as $.774\times10^{-3}$
   The mantissa is 774, the exponent is –3 and the sign of the number is positive. Therefore we can write the number in the 8–digit register as:

+	7	7	4	0	-	0	3

Now consider the numbers, 1,234,562,328 and 0.00084325667. It would not be possible to store or work with either of these numbers since they both contain more than 8 significant digits. Representing them as floating–point numbers doesn't help, i.e.

$$1{,}234{,}562{,}328 = .1234562328 * 10^{10}$$
$$\text{and}$$
$$0.0001432567 = .1432567 * 10^{-3}$$

Neither of the above two numbers can be represented exactly using our 8–digit computer. How do we work around this problem? In these cases we are forced to use approximations. Thus, we need to thoroughly understand what the potential implications are.

Objective B: Approximation and Error

Because of the limitations of a computer to accurately represent all numbers, we can generate errors in our calculations. In the previous section, we discussed the importance of floating–point numbers when using computers.

However, even floating–point numbers have limitations. For example, our register cannot hold any number larger than

$$+9999+99$$

or smaller than

$$-9999+99$$

Furthermore, it cannot hold the number 37,892.5 in its "exact" form, since the mantissa has more significant digits than our register can hold. This condition is called **overflow**. Typical computers today have many more than 8 characters in a register, but eventually every computer can potentially have an overflow condition. When this happens an error message may be sent by the computer's operating system (for instance, "floating–point error".) If an error message is not sent the computer may make adjustments by either truncating or rounding the number. Obviously, if this occurs the user should be aware of it, otherwise problems can develop. Let's examine the types of approximations we can use.

We can approximate a number in one of two ways, the first is by truncation and the second is by rounding.

Truncation

To truncate a number at a particular decimal place, we simply delete all the digits to the right of that place.

Examples:

Fixed–Point

Number	Truncated at the decimal point	Truncated at the first decimal	Truncated at the second decimal	Truncated at the third decimal
179.345678	179	179.3	179.34	179.345
41.77796	41	41.7	41.77	41.777
−123.1246	−123	−123.1	−123.12	−123.124
13.9	13	13.9	13.90	13.900

Floating–Point

Number	Truncated one digit	Truncated two digits	Truncated three digits	Truncated four digits
$.54789*10^5$	$.5*10^5$	$.54*10^5$	$.547*10^5$	$.5478*10^5$
$-.33756*10^{-3}$	$.3*10^{-3}$	$.33*10^{-3}$	$.337*10^{-3}$	$.3375*10^{-3}$
$.2929436*10^8$	$.2*10^8$	$.29*10^8$	$.292*10^8$	$.2929*10^8$
$.119*10^{-9}$	$.1*10^{-9}$	$.11*10^{-9}$	$.119*10^{-9}$	$.1190*10^{-9}$

Rounding

To round a number to a particular decimal place, you must consider the next digit to the right of the decimal place where you are rounding.

If the next digit to the right of where you are rounding is:
a. 0, 1, 2, 3, or 4, then the original digit at the decimal place where you are rounding is left unchanged and all the following digits are deleted;

b. 5, 6, 7, 8, or 9, then the original digit at the decimal place where you are rounding is increased by 1 (carrying if necessary) and all the following digits are deleted.

In rounding a number you often look at what are called **significant digits.** Significant digits are the number of digits you consider necessary for, or important in, the calculation. In a fixed–point number the number of significant digits can either mean the number of total digits or it is sometimes referred to as digits to the right of the decimal place.

Care must be taken here. For example, in the problem above, we showed four different approximations for the number 179.345678. They were 179, 179.3, 179.34 and 179.345. These approximations can be viewed in one of two ways:

- 179 could be viewed as having 3 significant digits or no significant digits to the right of the decimal point.

- 179.3 could be viewed as having 4 significant digits or 1 significant digit to the right of the decimal point.

- 179.34 could be viewed as having 5 significant digits or 2 significant digits to the right of the decimal point.

- 179.345 could be viewed as having 6 significant digits or 3 significant digits to the right of the decimal point.

For a floating–point number the number of significant digits is just the total number of digits that you are rounding to.

Examples:
Fixed–Point

Number	Rounded to one digit	Rounded to two digits	Rounded to three digits	Rounded to four digits
179.345678	200	180	179	179.3
41.77796	40	42	41.8	41.78
−123.1246	−100	−120	−123	−123.1
13.95	10	14	14.0	13.95

Floating–Point

Number	Rounded to one digit	Rounded to two digits	Rounded to three digits	Rounded to four digits
$.54789*10^5$	$.5*10^5$	$.55*10^5$	$.548*10^5$	$.5479*10^5$
$-.33756*10^{-3}$	$-.3*10^{-3}$	$-.34*10^{-3}$	$-.338*10^{-3}$	$-.3376*10^{-3}$
$.2991436*10^8$	$.3*10^8$	$.30*10^8$	$.299*10^8$	$.2991*10^8$
$.119*10^{-9}$	$.1*10^{-9}$	$.12*10^{-9}$	$.119*10^{-9}$	$.1190*10^{-9}$

The processes of truncating and rounding yield numbers that are approximations to the original numbers. This introduces an error into the calculation process. Depending upon the algorithm, this may or may not have a significant impact on the problem. The discussion of such errors is extensive and can be quite difficult to analyze. For the most part this is beyond the scope of this course, but we shall introduce the concept of absolute and relative error to illustrate the potential problem. This is something you need to be aware of when working with computers.

Objective C: Absolute and Relative Error (Optional)

We now introduce the concepts of absolute error and relative error, and apply them to a few examples of common errors caused by using approximation in computers.

Definition: Given a number **Num** and its approximation **App**, we define the absolute error (**AbsErr**) by the difference in the number and its approximation, i.e.

$$AbsErr = Num - App$$

and, the relative error (**RelErr**) by the ratio of the absolute error and the number we are approximating, i.e.

$$RelErr = AbsErr/Num = (Num - App)/Num$$

Usually the relative error is expressed as a percentage.

Of these two quantities, the relative error is generally the better measure of how the approximation impacts the result.

For example, if 100,000 is approximated by 99,999 and 10 is approximated by 9, then the absolute error is 1 for each, but the relative error is

1/100,000=0.00001=0.001% for the first case, and

1/10=0.1=10% for the second case.

Obviously the relative error shows that the second approximation has a much more significant impact, although the absolute error is the same for both examples.

To illustrate further, let's take a look at the absolute and relative error for the following examples of fixed–point numbers:

Truncation Error:

If 72.5689 is truncated at the first decimal place, then

AbsErr = 72.5689–72.5 = 0.0689, and

RelErr = (0.0689/72.5689)*100 = 0.095%

Rounding Error:

If 72.5689 is rounded to the first decimal place, then

AbsErr = 72.5689–72.6 = –0.0311, and

RelErr = (–.0311/72.5689)*100 = –0.043%

Here we see that the relative error is negative. This happens because by rounding up the approximation was larger than the actual value, thus we are overestimating the result by 0.043%. A negative relative or absolute error tells us the estimate or approximation is larger than the actual value, while a positive error, tells us it is less than the actual value.

Now consider the same examples as floating–point numbers:

Truncation Error:

If $.725689*10^2$ is truncated at the third decimal place, then

AbsErr $= .725689*10^2 - .725*10^2 = .689*10^{-1}$, and

RelErr $= (.689*10^{-1}/.725689*10^2)*100 = 0.095\%$

Rounding Error:

If $.725689$ is rounded to the third decimal place, then

AbsErr $= .725689*10^2 - .726*10^2 = -.311*10^{-1}$, and

RelErr $= (-.311*10^{-1}/.725689*10^2)*100 = -0.043\%$

It's important to note that the relative error is the same in the fixed point form as it is for the same number in floating point form.

Objective D: Conversion, Accumulation, and Subtraction Error (Optional Section)

Because of the finite size of a computer register and the fact that numbers must be approximated on a computer, several different types of errors can occur.

Conversion Error

The first type of error we encounter is the **conversion** error. Conversion error occurs when we must convert an irrational number or a rational number whose decimal representation does not terminate, into a decimal number that can be stored in a computer with a finite number of bits.

Consider the following examples:

Example 1: Compute the absolute and relative error when the fraction $1/3 = 0.333\ldots$ is stored as a floating–point number with a mantissa truncated to four digits.

   Solution: We begin with the exact value, **Num**, and the 4–digit approximation, **App**, and then compute the **AbsErr** and the **RelErr** using the formula from above.

$$\text{Num} = \frac{1}{3}$$
$$\text{App} = .3333\text{E}0$$

$$\text{AbsErr} = \text{Num} - \text{App} = \frac{1}{3} - .3333\text{E}0 = .0000333\ldots$$

$$\text{RelErr} = \frac{\text{AbsErr}}{\text{Num}} = \frac{.0000333\ldots}{\frac{1}{3}} = .0001 = .01\%$$

Note: In example 1 we saw the notation E0 for the first time. This is how many computers will represent the number 10^0,. In example 2 we have the number E1 which is the number 10^1. Any time we show the number EN, this will mean 10^N.

Example 2: Compute the absolute and relative error when the number $\sqrt{2} = 1.4142135\ldots$ is stored as a floating–point number with a mantissa truncated to four digits.

Solution: We begin with the exact value, **Num**, and the 4–digit approximation, **App**, and then compute the **AbsErr** and the **RelErr** using the formula from above.

$$\text{Num} = \sqrt{2} = 1.41421356237\ldots$$
$$\text{App} = .1414\text{E}01$$

$$\text{AbsErr} = .0002135623\ldots$$
$$\text{RelErr} = \frac{\text{AbsErr}}{\text{Num}} = \frac{.0002135623\ldots}{1.41421356237\ldots} \approx .000151 = .0151\%$$

Accumulation Error

```
 .3333                       1.666          2.998
+.3333                      +.3333         +.3333
 .6666                       1.999          3.331
+.3333                      +.3333         +.3333
 .9999                       2.332          3.664
+.3333                      +.3333         +.3333
 1.333    ← Mantissa         2.665          3.997
+.3333       truncated      +.3333
 1.666    ← To 4 digits      2.998
```

Conversion errors are typically small. However, quite often we have to perform repeated calculations where we must convert numbers many times. The error then accumulates. This is called **accumulation** error.

For example, what if we had to add 1/3 to itself 12 times in a calculation with a mantissa of only 4 digits as in the examples above. Then,

$$\mathbf{Num} = \frac{1}{3} * 12 = 4$$

$$\mathbf{App} = 3.997$$

$$\mathbf{AbsErr} = \mathbf{Num} - \mathbf{App} = 4 - 3.997 = .003$$

$$\mathbf{RelErr} = \frac{\mathbf{AbsErr}}{\mathbf{Num}} = \frac{.003}{4} = .00075 = .075\%$$

We should note that this error is 7.5 times larger than the error caused by a single conversion as in the example above. Thus, accumulation can have a significant impact on the result. Imagine if you had to add such numbers say a hundred or even a thousand or more times. The accumulation error can grow quite large.

Subtraction Error

The last type of error we consider occurs when we have to subtract two numbers that are relatively close to one another.

Consider the two numbers

.62194E04 and .62186E04

The actual difference is
  .62194E04
 −.62186E04
  .00008E04

whereas if the computer truncates to four (4) mantissa digits, then it computes

  .6219E04
 −.6218E04
  .0001E04

Thus,

$$\begin{aligned} \text{Num} &= .00008\text{E}04 \\ -\ \text{App} &= .0001\ \text{E}04 \\ \hline \text{AbsErr} &= -.00002\text{E}04 \end{aligned}$$

$$\textbf{RelErr} = \frac{\textbf{AbsErr}}{\textbf{Num}} = \frac{-.00002\text{E}04}{.00008\text{E}04} = \left(\frac{-.00002}{.00008}\right) \div \text{E}(4-4) = -.25\text{E}0 = -25\%$$

The relative error is an astonishing 25%!

In summary, we have looked at five different types of errors in this section. They include rounding, truncation, conversion, accumulation, and subtraction errors. We have also shown that there are two ways to measure such errors. We can either measure the absolute error, which just looks at the overall error, or the relative error, which looks at the error relative to the size of the answer. The relative error gives a better measure of how big an impact the error will have, because it tells us how big the change is as a percent of the size of the answer we are working with. Also, we need to remember that the two most troublesome errors are subtraction and accumulation errors, since they can have a large impact on our calculations and we must be aware of it.

EXERCISES 2.2

Given a computer with a register that can hold 10 digits, show how the following numbers would be stored in a computer:

1. −9,459,300,000,000,000
2. 0.000 000 117 852
3. 1,235,670
4. −0.000 000 629 44

Given a computer with a register that can hold 8 digits, show how the following numbers would be stored:

5. 645,300,000,000
6. −0.000 000 000 121 300
7. −5,396
8. 0.000 000 383 20

9. What is computer overflow?

Given the following charts, fill in the missing data.

Fixed–Point Number	Rounded to one digit	Rounded to two digits	Rounded to three digits	Rounded to four digits
10. 789.124				
11. −35.23897				
12. 13.77595				
13. −1,024.092				

Floating–Point Number	Rounded to one digit	Rounded to two digits	Rounded to three digits	Rounded to four digits
14. $.12399*10^{-3}$				
15. $-.9875*10^{7}$				
16. $.23556*10^{6}$				
17. $.2493*10^{-11}$				

Given the following charts, fill in the missing data.

Fixed–Point Number	Truncated to one digit	Truncated to two digits	Truncated to three digits	Truncated to four digits
18. 789.124				

19. −35.23897 _____
20. 13.77595 _____
21. −1,024.092 _____

	Floating–Point Number	Truncated to one digit	Truncated to two digits	Truncated to three digits	Truncated to four digits
22.	.12399*10⁻³				
23.	−.9875*10⁷				
24.	.23556*10⁶				
25.	.2493*10⁻¹¹				

The exercises that follow are from optional sections within the chapter.

Compute the Absolute and Relative errors of the following:
26. 124.395 truncated at the second decimal place
27. 124.395 rounded at the second decimal place
28. 0.0195 truncated at the third decimal place
29. 0.0195 rounded at the third decimal place
30. 46.496 truncated at the first decimal place
31. 46.496 rounded at the first decimal place

32. Compute the absolute and relative conversion errors involved with storing the number 2/3 in a computer that truncates after the second decimal place.

33. Compute the absolute and relative conversion errors involved with storing the number $\sqrt{3}$ in a computer that truncates after the second decimal place.

34. Compute the absolute and relative conversion errors involved with storing the number 2/3 in a computer that rounds after the second decimal place.

35. Compute the absolute and relative conversion errors involved with storing the number $\sqrt{3}$ in a computer that rounds after the second decimal place.

36. Compute the absolute and relative accumulation errors that result if we add 2/3 to itself 10 times on a computer that can only store a mantissa of 3 digits.

37. Compute the absolute and relative accumulation errors that result if we add 1/6 to itself 15 times on a computer that can only store a mantissa of 4 digits.

38. Consider the two numbers: .1129E−5 and .1124E−5, compute the absolute and relative errors that occur when the second number is subtracted from the first on a computer that truncates to three and then four mantissa digits.

39. Consider the two numbers: .9769E+9 and .9771E+9, compute the absolute and relative errors that occur when the second number is subtracted from the first on a computer that truncates to three and then four mantissa digits.

COMPUTER ACTIVITY

1. Write a computer program that will input a number and then output the absolute and relative errors of both rounding and truncating the inputted number to two decimal places.

2. Write a computer program that will input a rational number (p/q) and then output the absolute and relative errors obtained when converting this number to a decimal number by truncating and rounding on a computer with only 4 decimal places. Use the numbers 1/6, 1/9 and 21/22 as input examples, but make the program generic for any rational number.

3. Write a computer program that will input a number and then take its square root. The program should then output the absolute and relative errors obtained when converting this radical number to a decimal number by truncating and rounding on a computer with only 1 decimal place. Use the numbers 3, 7, and 11 as input examples, but make the program generic for any positive number.

Section 2.3: Decimal, Binary, Octal and Hexadecimal Numbers

As we discussed in Section 2.1, a computer essentially processes data in bits and bytes, which are represented by binary numbers. Before we can understand the binary number system, a thorough understanding of the base 10 or decimal number system is required.

First, consider the decimal number 4,285. We know that we can write this number in "expanded form" as:

$$4,285 = 4*1000 + 2*100 + 8*10 + 5*1$$
$$= 4*10^3 + 2*10^2 + 8*10^1 + 5*10^0$$

Notice that the terms 10^3, 10^2, 10^1 and 10^0 are all powers of 10. We read the number as four thousand two hundred eighty five. That means this number has 4 thousands, 2 hundreds, 8 tens and 5 ones.

In base 10, there are 10 possible digits (symbols) that can occupy any position in the number. The digits are 0, 1, 2, 3, 4, 5, 6, 7, 8 and 9. This is a positional number system, where the location of a digit in a number corresponds to a specific "place value." The value of the place the digit is located in, is equal to 10 raised to an integer power. The value of the power is determined by the position of the digit in the number.

Here is a table showing the place values of the digits in the number 4,285

Decimal	4	2	8	5
Place value	10^3=1000	10^2=100	10^1=10	10^0=1

Since the number we are looking at is in base 10, the bottom row represents the place values of the digits above it. These entries are just 10 raised to the indicated exponent. Also note that in any position, the digit can only go as high as 9.

The two most important features we should understand about our base 10 number system, is that the number of symbols we need to represent digits, is equal to the base (10), and the place value of a digit is just the base (10) raised to a power that corresponds to the position of the digit in the number. This is what defines a number system.

Historically it's assumed that we use the base 10 number system because we have 10 fingers. There are, however, other number systems. Number systems are identified by using either their base value or by using their associated name. In the table below we provide a list of the various bases along with their corresponding names.

Base Value	Base Name	Base Value	Base Name
2	Binary (binal)	16	Hexadecimal
3	Trinary (trial)	17	Septendecimal
4	Quarternary (quartal)	18	Decennoctal
5	Quinary (quintal)	19	Decennoval
6	Sextal	20	Vigesimal
7	Septimal	30	Trigesimal
8	Octal	40	Quadragesimal
9	Nonal	50	Quinquagesimal
10	Decimal	60	Sexagesimal
11	Undecimal	70	Septagesimal
12	Duodecimal	80	Octagesimal
13	Tridecimal	90	Nonagesimal
14	Quaturodecimal	100	Centesimal
15	Quindecimal	1000	Millesimal

In this section we shall focus on three number systems that are very important in computer studies: they are the binary, octal, and hexadecimal number systems. We will first consider how they are formed and then show how to convert them into their more familiar decimal representations. Finally, we shall discuss how to convert any decimal number to another base and how to perform arithmetic operations using these different number systems.

Objective A: Converting Binary, Octal, and Hexadecimal to Decimal

Recall that the decimal system has 10 symbols from which to compose numbers and has place values based upon powers of 10. To consider other base number systems we only need to change two things: the number of symbols and the value of the digit location in our number. Thus, we can extend what we've shown above for decimal numbers to any positive base.

The possible digits for: base 2 are 0, 1
base 3 are 0, 1, 2
base 4 are 0, 1, 2, 3
base 5 are 0, 1, 2, 3, 4
etc.

A base "n" number system has "n" distinct symbols that represent digits and 0 (zero) is always one of the digits.

The first system we shall introduce is the binary system. It is critical that you thoroughly understand this system if you are going to be working with computers.

Binary Numbers

A binary number is a base 2 number. Consequently, it has only 2 symbols to use as digits, 0 and 1. You cannot have a 2 or larger in any digit of a binary number. So how do we write binary numbers?

Let's start with some simple counting in the binary system. An example of counting from 0 (zero) to 11 (eleven) is shown in the table below.

Decimal	Binary	Binary (shown as addition)	Decimal equivalent (shown as addition)
0	0	0	0
1	1	0+1=1	0+1=1
2	10	1+1=10 (note: two places)	1+1=2
3	11	10+1=11	2+1=3
4	100	11+1=100	3+1=4
5	101	100+1=101	4+1=5
6	110	101+1=110	5+1=6
7	111	110+1=111	6+1=7
8	1000	111+1=1000	7+1=8
9	1001	1000+1=1001	8+1=9
10	1010	1001+1=1010	9+1=10 (note: two places)
11	1011	1010+1=1011	10+1=11

We should note from the above chart, that in binary, 4 digits are needed to represent the decimal number 8. We also see a pattern developing as we count. In the right most position, if the value is a 0 it turns into a 1. If the value is a 1 it turns into 0 and you add 1 to the next position to the left and repeat. Also we need to point out that reading the name of a binary number is different than reading a decimal number. For example the binary number 111_2 is read one–one–one, not one hundred and eleven.

Example 1: What is the binary equivalent to the decimal number 9?

Solution: From the chart above we see that the decimal number 9 corresponds to the binary number 1001_2. Notice that to help identify that we are no longer in base 10, we have written a subscript to the right of the number. The number in this subscript is used to identify the base we are working in. In this case it is a base 2 number. Throughout this chapter we will use subscripts to identify the base, unless it is obvious which base we are working in. Whenever the base is not indicated, base 10 is the default base.

We will now show how we can convert any binary number into a more familiar decimal equivalent. Using the place value table that we used for decimals at the beginning of this section, we can construct a similar table for a binary number. Instead of raising 10 to a power for the place value, we now raise 2 to an appropriate exponent depending on its position in the binary number.

The table starts with the binary number we are converting from (top row), and ends with the decimal number we are converting to (bottom row). The binary number is placed on the top row. The decimal equivalent is obtained by multiplying the binary digit of the first row by the binary place value in the second row. Summing the resulting numbers in the third row gives the decimal equivalent of the binary number.

For example, according to the table above, 7 in base 10 is equivalent to 111 in base 2. To verify this, let's construct a binary table.

Binary	1	1	1
Place value	$2^2=4$	$2^1=2$	$2^0=1$
Decimal Equivalent	4	2	1

To get the equivalent decimal number, we sum up the numbers in the bottom row, i.e.,

$$4+2+1 = 7$$

Instead of using a table, we could also write our binary number in expanded notation, i.e.,

$$\begin{aligned}111_2 &= 1*2^2 + 1*2^1 + 1*2^0 \\ &= 1*4 + 1*2 + 1*1 \\ &= 7_{10}\end{aligned}$$

Note: We have used a subscript here to represent the base our number is in. 7_{10} is the number 7 in a base 10 (decimal) system.

We have shown that converting a binary number to a decimal is equivalent to writing the binary number in expanded form and then summing the final row in the place value table.

Example 2: Convert 101101_2 to base 10.

Solution: Since there are six digits in the number, we construct our table with 6 (six) place values. The first row contains our binary number. In the next row we set up the powers of two, starting from the right–most digit having a zero exponent, then work our way to the left–most digit. In this example, there are six digits, so the highest power of 2 is 5. We then write the decimal equivalent number for each of the powers. The final row is obtained by multiplying the number in the first row by the number in the second row. Then we sum the numbers in the last row of our place value table to obtain our decimal equivalent number.

Binary	1	0	1	1	0	1
Place value	$2^5=32$	$2^4=16$	$2^3=8$	$2^2=4$	$2^1=2$	$2^0=1$
Decimal Equivalent	32	0	8	4	0	1

Summing the bottom row we have:

$$32+8+4+1 = 45$$

We can also write our number directly using its expanded form as:

$$\begin{aligned}101101_2 &= 1*2^5 + 0*2^4 + 1*2^3 + 1*2^2 + 0*2^1 + 1*2^0 \\ &= 1*32 + 0*16 + 1*8 + 1*4 + 0*2 + 1*1 \\ &= 45_{10}\end{aligned}$$

Thus, $101101_2 = 45_{10}$

Example 3: Convert 1111001_2 to base 10.

Solution: There are seven digits in the number so we construct our table with 7 place values. We assign place values to the binary numbers as indicated in the table below and compute the expanded form of the number.

Binary	1	1	1	1	0	0	1
Place value	2^6=64	2^5=32	2^4=16	2^3=8	2^2=4	2^1=2	2^0=1
Decimal	64	32	16	8	0	0	1

Summing the bottom row we have:

$$64+32+16+8+1 = 121$$

We can also write our number directly using its expanded form as:

$$\begin{aligned}1111001_2 &= 1*2^6 + 1*2^5 + 1*2^4 + 1*2^3 + 0*2^2 + 0*2^1 + 1*2^0 \\ &= 1*64 + 1*32 + 1*16 + 1*8 + 0*4 + 0*2 + 1*1 \\ &= 121_{10}\end{aligned}$$

Thus, $1111001_2 = 121_{10}$

Octal Numbers

An octal is a base 8 number. This means there are 8 different symbols that can be used for digits. These are the digits 0, 1, 2, 3, 4, 5, 6 and 7. Furthermore, each place value of an octal number corresponds to 8 raised to an integer power.

Before we show how to convert an octal number to a decimal, let us compare numbers in base 10, base 2 and base 8 by counting upward from zero.

decimal (base10)	binary (base 2)	Octal (base8)
0	0	0
1	1	1
2	10	2
3	11	3
4	100	4
5	101	5

decimal (base10)	binary (base 2)	Octal (base8)
6	110	6
7	111	7
8	1000	10
9	1001	11
10	1010	12
11	1011	13
12	1100	14

Example 4: What is the octal equivalent of the decimal number 11?

Solution: From the chart above, we see that the number 11 corresponds to the octal number 13_8.

Example 5: What is the octal equivalent of the binary number 1000_2?

Solution: From the chart above, we see that the number 1000_2 corresponds to the octal number 10_8.

To convert an octal number to a decimal we follow the same procedure we used for binaries. We illustrate this using an example.

Example 6: Convert 324_8 to base 10

Solution: We follow the same process as above, only now we change the base value to 8. We begin with our table, where the number we are converting from is placed in the top row and the decimal equivalent is obtained from the bottom row. The center row is simply the place value of the octal digit. We then multiply the octal digit by the place value to get the decimal equivalent for that place value position and write this in the bottom row.

Octal	3	2	4
Place value	$8^2=64$	$8^1=8$	$8^0=1$
Decimal Equivalent	$3*64=192$	$2*8=16$	$4*1=4$

Finally, we sum the bottom row to obtain our answer.

$$192+16+4 = 212_{10}$$

Alternatively we can write the number in expanded form, i.e.

$$324_8 = 3*8^2 + 2*8^1 + 4*8^0$$
$$= 3*64 + 2*8 + 4*1$$
$$= 212_{10}$$

Hexadecimal Numbers

A hexadecimal number is a base 16 number, which means there are 16 different digits that can be used in any one position. We obtain these digits by starting with our standard symbols for whole numbers, namely: 0, 1, 2, 3, 4, 5, 6, 7, 8 and 9. However, these are only 10 numbers (digits). In base 16 we need a total of 16 different symbols. The next counting number 10, is not a single digit (since it physically occupies two places for the one and the zero), and this cannot happen. Instead we use letters to designate the six remaining symbols. They are A for 10, B for 11, C for 12, D for 13, E for 14, and F for 15. Note that 15 is the last numerical value and is exactly one less than the base value of 16.

Now, before we try to convert a hexadecimal number to base 10, let us compare numbers in base 10, base 2 and base 16 by counting upward from zero.

decimal (base10)	binary (base 2)	hexadecimal (base16)	decimal (base10)	binary (base 2)	hexadecimal (base16)
0	0	0	10	1010	A
1	1	1	11	1011	B
2	10	2	12	1100	C
3	11	3	13	1101	D
4	100	4	14	1110	E
5	101	5	15	1111	F
6	110	6	16	10000	10
7	111	7	17	10001	11
8	1000	8	18	10010	12
9	1001	9			

Example 7: What is the hexadecimal equivalent to the decimal number 17?

Solution: From the chart above we see that the number 17 in base 10 corresponds to the hexadecimal number 11_{16}, i.e. $17_{10} = 11_{16}$. We should note that when reading this 11_{16} does not represent 11 (eleven) as we are used to reading it, it should be read as one–one in base 16.

Let us now show how we take a hexadecimal number and rewrite it as a decimal. We follow a similar procedure we developed for changing binaries and octals to decimals. Only now we are in a base 16 and not a base 2 or base 8 number system.

We will now show how to convert a hexadecimal number to decimal using an example.

Example 8: Convert the hexadecimal number CAB_{16} to a decimal number

Solution: Since there are three digits in the number, let us construct our table with 3 place values. The first row contains our hexadecimal number. In the next row we set up the powers of sixteen, starting from the right–most digit having a zero exponent, then work our way to the left–most digit. In this example, there are three digits, so the highest power of 16 is 2. The third row is just the decimal equivalent value of the particular hexadecimal digit. The final row is obtained by multiplying the number in the fourth row by the number in the third row. Then we simply sum the numbers in the last row of our place value table.

Hexadecimal	C	A	B
Place value	$16^2=256$	$16^1=16$	$16^0=1$
Digit value	C=12	A=10	B=11
Decimal Equivalent	$12*256=3072$	$10*16=160$	$11*1=11$

Summing the bottom row we obtain:

$$3072+160+11 = 3243$$

We can also write our number directly using its expanded form as:
$$CAB_{16} = C*16^2 + A*16^1 + B*16^0$$
$$= 12*256 + 10*16 + 11*1$$
$$= 3243_{10}$$

Thus, $CAB_{16} = 3243_{10}$

Example 9: Convert $2A5_{16}$ to a decimal

Solution: We follow the same procedure as above, and begin with our table.

Hexadecimal	2	A=10	5
Place value	$16^2=256$	$16^1=16$	$16^0=1$
Decimal Equivalent	$2\times 256 = 512$	$10\times 16 = 160$	$5\times 1 = 5$

Summing the bottom row we have:

$$512+160+5 = 677$$

Now, let us write the conversion in expanded form:

$$\begin{aligned} 2A5_{16} &= 2\times 16^2 + A\times 16^1 + 5\times 16^0 \\ &= 2\times 256 + 10\times 16 + 5\times 1 \\ &= 677_{10} \end{aligned}$$

Objective B: Converting Decimal to Binary and Hexadecimal

Now that we know how to convert a number in a base other than 10 to base 10, how do we convert a number in base 10 to another base? For example, what if we wanted to convert the decimal number 99 to a binary or a hexadecimal number. How would we go about doing this?

There are two possible ways of doing the conversion. The first method we'll show involves our familiar table, but this time instead of adding to get the number, we'll subtract to find the number. The second approach involves division and a remainder term.

Converting Decimal to Binary

Method 1: To convert 99_{10} to its binary equivalent, let's first create a binary place value table.

2^7	2^6	2^5	2^4	2^3	2^2	2^1	2^0
128	64	32	16	8	4	2	1

We start by looking at the place values from the left of the table to the right. Looking at the table below we notice that the eighth place value from the right is $2^7 =128$. This number is greater than 99 so it would be too large. So we know that the binary number that we are looking for cannot have a 1 (one) in the eighth place value, otherwise the number would have to be 128 or larger.

The next smallest value is $2^6=64$, which is a number that is less than 99. So our number must contain a digit in the seventh binary place. That means our binary number must be seven digits long.

To do the conversion we simply subtract the largest place value from our number and rewrite the number as the sum of two new numbers. More specifically, 99–64 = 35, thus, 99 can be rewritten as 64 + 35.

Then we notice that $2^5=32$ is less than the remainder, 35. Thus 35 can be rewritten as 32+3., since 35–32 = 3. Now, since 3 is less than 16, 8, and 4, the number does not contain any of these place values. For these digits there will be all 0's in those positions. Continuing on we see that 2 is less than 3 so we can rewrite 3 as 2+1.

Thus, we have rewritten 99 as 64+32+2+1, i.e., we have **decomposed** 99 into a sum of the place values of a binary number. Putting this into a table we see that:

Binary Place value	128	64	32	16	8	4	2	1
Binary Digits	0	1	1	0	0	0	1	1

We recall from above that to obtain the decimal equivalent, we multiplied the top row (binary place value) by the binary digit, to get the decimal equivalent of that place value. To get the binary equivalent, we look for the binary digit (a 1 or 0) that when multiplied by the binary place value, gives us the decimal decomposition.

Thus, the binary equivalent of 99_{10} is 1100011_2, since 1100011_2 =64+32+2+1=99

Another way to view this is that 99 can be decomposed into one group of 64, one group of 32, no group of 16, no group of 8, no group of 4, one group of 2 and one group of 1.

Method 2: To convert 99_{10} to a binary number, ( )$_2$, we start by dividing 99 by 2 (the base we are converting to) until the quotient is 0. At the same time we keep track of the remainder term.

Notice that when we divide 99 by 2, the quotient is 49 with a remainder of 1. We place the remainder in its own column, as shown below. Then we continue and divide 49 by 2, placing the quotient under the previous quotient and its remainder next to it under the remainder column.

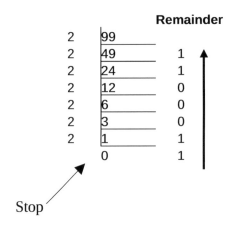

Stop

We continue to do this until the quotient is 0. Also notice that even if the remainder is 0, we put it down. We knew from our previous work that the number 99 requires 7 digits in binary. Notice that we have 7 remainders from our division process.

The remainders written from the bottom up will be the binary equivalent number, i.e., $99_{10} = 1100011_2$. This is the same number we obtained using the table approach.

Example 1: Convert the decimal number 239 to a binary number.

Solution: **Method 1**: Consider the table of binary powers:

2^8	2^7	2^6	2^5	2^4	2^3	2^2	2^1	2^0
256	128	64	32	16	8	4	2	1

By comparing 239 to the chart above we see that it is less than 256 and greater than 128, therefore our binary number will have 8–digits and not 9. The first digit will give us a value of 128 which when subtracted from 239 yields:
$$239-128 = 111$$

Continuing in the same way we see that 111 is less than 128, but greater than 64, therefore, it contains one 64. Now subtract 64 from 111 to obtain:

$$111-64 = 47$$
Now 47 is less than 64 but greater than 32, so there is a 32 in the number.

$$47-32 = 15$$

15 is less than 16 but greater than 8 so there are no 16's and one 8 in the number.

$$15-8 = 7$$

7 is greater than 4, so there is one 4 and when we subtract 4 we have a remainder of 3, which is one two and one 1. Thus we've decomposed 239 into the following sum:

$$239 = 128+64+32+8+4+2+1$$

Putting this in a table we have:

Binary Place value	128	64	32	16	8	4	2	1
Binary Digits	1	1	1	0	1	1	1	1

Therefore, the binary equivalent of 239 is 11101111_2, since

$$11101111_2 = 128+64+32+8+4+2+1 = 239$$

Thus, 239 can be written in expanded form as:

$$239 = 1*128+1*64+1*32+0*16+1*8+1*4+1*2+1*1$$
$$= 1 \quad 1 \quad 1 \quad 0 \quad 1 \quad 1 \quad 1 \quad 1$$
$$= 11101111_2$$

Again, we can also say that 239 has one group of 128's with 111 left over. Then there is one group of 64's in 111 with 47 left over. There is one group of 32's in 47 with 15 left over. There is one group of 8's in 15 with 7 left over and there is one group of 4's in 7 with 3 left over. Finally, there is one group of 2's in 3 with 1 left over.

Method 2: To convert 239 base 10 to a binary number we use continuous division by 2 (the base we are converting to) until the quotient is 0. At the same time we keep track of the remainder term.

Notice that when we divide 239 by 2, the quotient is 119 with a remainder of 1. We place the remainder in its own column, as shown below. Then we continue and divide 119 by 2, placing the quotient under the previous quotient and its remainder next to it under the remainder column.

```
                    Remainder
2 | 239
2 | 119        1
2 | 59         1
2 | 29         1
2 | 14         1
2 | 7          0
2 | 3          1
2 | 1          1
    0          1
```

We continue to do this until the quotient is 0. Also notice that even if the remainder is 0, we put it down. We knew from our previous work that the number 239 requires 8 digits in binary. Notice that we have 8 remainders from our division process.

The remainders written from the bottom up form the binary equivalent, i.e., $239_{10} = 11101111_2$. This is the same number we obtained using the table approach.

Converting Decimal to Hexadecimal

We now look at the two different ways of converting a decimal into a hexadecimal number.

Method 1: We follow the same procedure as above, but now we are working with powers of 16 and utilizing 16 different digits. We illustrate this with an example and show how to convert the decimal number 99 into a hexadecimal number. We first determine how many digits the number 99_{10} has in base 16. Consider the place value table below for a hexadecimal number:

16^2	16^1	16^0
256	16	1

From the table above, we know that the hexadecimal number we are looking for has only two digits. There is no group of 256 in 99, since 256 is greater than 99. Moving on to the next place value to the right (16), we simply divide 99 by 16, which is 6 with a remainder of 3. That means there are 6 sixteen's in 99 plus three one's. We can rewrite 99 in the following way:

$$99 = 6 \times 16 + 3 \times 1$$

Putting this decomposition into a table we have:

Hexadecimal Place value	16	1
Hexadecimal Digits	6	3

Therefore, the hexadecimal equivalent of 99 is 63_{16}, since

$$63_{16} = 96 + 3 = 6*16 + 3*1 = 99_{10}$$

Thus, 99 can be written in expanded form as:
$$99 = 6*16 + 3*1$$
$$= 63_{16}$$

Method 2: Following the alternative method we used for converting a number in base 10 to binary, we can do the same in converting a number in base 10 to base 16 (hexadecimal). The difference is now we divide by 16, which is the base we want to convert to, instead of 2.

```
                     Remainder
        16  │ 99
        16  │ 6         3   ↑
              0         6   │
```

Divide 99 by 16 (the base), place the quotient under it and the remainder under the remainder column. Then divide 6 by 16, which is 0 with a remainder of 6. The answer is $99_{10} = 63_{16}$.

Example 2: Convert the number 339_{10} to a hexadecimal number.

Solution: **Method 1**: Consider the table of hexadecimal powers:

16^3	16^2	16^1	16^0
4096	256	16	1

By comparing 339 to the chart above we see that it is less than 4096 and greater than 256, therefore our hexadecimal number will have 3 digits and not 4. The first digit will give us a value of 256 which when subtracted from 339 yields:

$$339 - 256 = 83$$

Continuing in the same way we see that 83 is less than 256, but greater than 16, therefore, it contains one 16. Now subtract 16 from 83 to obtain:

$$83 - 16 = 67$$

Now 67 is still greater than 16, therefore, there are more groups of 16 in 83. In fact, 16 goes into 83 five times with a remainder of 3, i.e.,

$$83 - 5*16 = 83 - 80 = 3.$$

Thus we've decomposed 339 into the following sum:

$$339 = (1*256) + (5*16) + (1*3)$$

Putting this in a table we have:

Hexadecimal Place value	256	16	1
Hexadecimal Digits	1	5	3
Decimal Decomposition	256 = 1*256	80 = 5*16	3

Therefore, the hexadecimal equivalent of 339 is 153_{16}, since

$$153_{16} = 256 + 80 + 3 = 1*256 + 5*16 + 3*1 = 339_{10}$$

Thus, 339 can be written in expanded form as:

$$339 = 1*256 + 5*16 + 3*1$$
$$= 153_{16}$$

Method 2: To convert 339_{10} to a hexadecimal number we start by dividing 339 by 16 (the base we are converting to) until the quotient is 0. At the same time we keep track of the remainder term.

Notice that when we divide 339 by 16, the quotient is 21 with a remainder of 3. We place the remainder in its own column, as shown below. Then we continue and divide 21 by 16, placing the quotient under the previous quotient and its remainder next to it under the remainder column.

```
                    Remainder
        16 | 339
        16 |  21       3  ↑
        16 |   1       5  |
              0        1
```

We continue to do this until the quotient is 0.

The remainders written from the bottom up will be the hexadecimal equivalent number, i.e. $339_{10} = 153_{16}$. This is the same number we obtained using the table approach.

Objective C: Converting Between Binary, Octal and Hexadecimal

We have shown how to convert from a decimal to a binary, octal, or hexadecimal number and then how to convert from a binary, octal, or a hexadecimal number to a decimal. Sometimes, however, we need to convert from a binary to a hexadecimal or octal or vice–versa. We now show how to convert between any two bases.

The process is fairly straightforward. We first convert our number to a decimal and then convert the decimal to the desired base.

For example, to convert a binary number into a hexadecimal, we'd first convert the binary into its decimal equivalent, and then convert the decimal into a hex number. We'll illustrate this by using examples:

Example 1: Convert $1000\ 1111\ 1100_2$ to base 16.

Solution: We start by converting $1000\ 1111\ 1100_2$ into a decimal number. Using the process from above we can show that:

$$1000\ 1111\ 1100_2 = 2{,}300_{10}$$

Next, we convert $2{,}300_{10}$ into a hex number by the procedure from above to obtain:

$$2{,}300_{10} = 8FC_{16}$$

Example 2: Convert 175_8 to binary.

Solution: We start by converting 175_8 into a decimal number. Using the process from above we can show that:

$$175_8 = 125_{10}$$

Next, we convert 125_{10} into a binary number by the procedure from above to obtain:

$$125_{10} = 1111101_2$$

Example 3: Convert 7765_8 to hex.

Solution: We start by converting 7765_8 into a decimal number. Using the process from above we can show that:

$$7765_8 = 4085_{10}$$

Next, we convert 4085_{10} into a hex number by the procedure from above to obtain:

$$4085_{10} = FF5_{16}$$

This procedure will work fine, but it turns out that for binaries, octals, and hexadecimals there is a special relationship. This is due to the fact that both 8 and 16 are evenly divisible by 2 (2 goes into 8, four times and 2 goes into 16, eight times.) As a result, there is a shortcut for doing conversions between binaries and both octals and hexadecimals.

Now let's look at the connection between binary and hexadecimal numbers. In the example above 99 is equivalent to

$$110\ 0011 \quad \text{in base 2}$$
$$6 \quad 3 \quad \text{in base 16}$$

Grouping the binary number in groups of 4 digits beginning from the right, you can see that each group of 4 binary digits is equivalent to one hexadecimal digit, i.e. $0011_2 = 3_{16}$ and $110_2 = 6_{16}$. The 110_2 term could have been written as 0110_2 to be consistent.

Now that we know the relationship between binary and hexadecimal numbers, it is easy to convert from one to the other. Let's try one and convert a binary to a hexadecimal.

We start with a binary number $0110\ 1001\ 1010_2$. Every 4 binary digits beginning from the right, makes up one hexadecimal digit. Thus, we divide our binary number into chunks of 4 digits and convert each 4 digit binary into the equivalent hexadecimal, as shown in the table below.

0110	1001	1010
6	9	A

$$0110\ 1001\ 1010_2 = 69A_{16}$$

Example 4: Convert $100\ 1110\ 0111\ 1010_2$ to base 16.

Solution: Using the table, divide the binary number in groups of 4 beginning with the right and determine the hexadecimal value for each group. The resulting number using a table is,

100	1110	0111	1010
4	14	7	10
4	E	7	A

or,

$$100\ 1110\ 0111\ 1010_2 = 4\ \ 14\ \ 7\ \ 10$$
$$= 4\ \ E\ \ 7\ \ A$$
$$= 4E7A_{16}$$

Example 5: Convert $1000\ 1111\ 1100_2$ to base 16.

Solution: Using a table, we divide the binary number in groups of 4, beginning with the right. Then we determine the hexadecimal value for each group of 4 binary numbers. The resulting number is the converted value, i.e.

1000	1111	1100
8	15	12
8	F	C

or

$$1000\ 1111\ 1100_2 = 8\ \ 15\ \ 12$$
$$= 8\ \ F\ \ C$$
$$= 8FC_{16}$$

Notice that we obtained the same result as in Example 1, but more quickly.

We can also convert a hexadecimal number to binary using the same method above.

Example 6: Convert $FF3A_{16}$ to binary.

Solution: Knowing that each hexadecimal digit can be expanded to 4 binary digits, we can write a table where the top row contains the hex number and the bottom row is the binary equivalent to the hex number above it, i.e.

F	F	3	A
1111	1111	0011	1010

or

$$FF3A_{16} = 1111\ 1111\ 0011\ 1010_2$$

The same idea works for the binaries and octals as well. Only now we use 3 binary digits for 1 octal digit, i.e.

Example 7: Convert $111\ 101\ 100_2$ to octal.

Solution: We start by constructing a table, but this time with 3 binary digits in each column.

111	101	100
7	5	4

or, $111\ 101\ 100_2 = 754_8$

EXERCISES 2.3

1. Count from 0 to 36 in the binary, octal, and hexadecimal number systems.

2. Express as a power of 2 the number of bits in a byte; assume there are 8 bits to a byte.

3. Express as a power of 2 the number of bits in 4 bytes.

4. What is the smallest number that is a power of 2 and is greater than or equal to
   a. 19

b. 48
   c. 113
   d. 429

5. What is the smallest number that is a power of 2 and is greater than or equal to
   a. 17
   b. 72
   c. 135
   d. 68

Convert the binary number (base 2) to decimal notation (base 10).
6. 1100
7. 10 1111
8. 10 0100
9. 1110 1001 0010
10. 10 0001 0110

Convert the hexadecimal numbers (base 16) to decimal form (base 10)
11. C00
12. 9872
13. ABCD
14. 38C5
15. 6E4C
16. A7

Convert the octal numbers (base 8) to decimal form (base 10)
17. 427
18. 1556
19. 216
20. 4422

Convert each of the following decimal numbers to binary (base 2), octal (base 8) and hexadecimal (base 6)
21. 37
22. 239
23. 16
24. 227
25. 55
26. 18

Section 2.4: Fractional Numbers

Up until this time we have only considered converting and representing whole numbers in different bases. We will now consider numbers that have a fractional part such as the decimal number 10.5. Just as we can represent fractional numbers using decimals, we can also represent these same numbers in other bases. In this section we show how to represent numbers with a fractional part in any base. To help us in understanding this concept, we shall first review the process for decimals.

Suppose we have a base 10 number that has a fractional part or "decimal," such as 123.45. We know that the place value of the digits placed to the left of the decimal point go in increasing powers of 10. The digit 3, to the left of the decimal point, is in the ones or 10^0 location, the next digit (2) is in the tens or 10^1 location and the next digit (1) is in the hundreds or 10^2 position. As we go further and further to the left we simply increase the exponent by 1. What are the place values of the digits to the right of the decimal point?

The 4 in the number stands for the number of tenths (1/10 or .1 or 10^{-1}) in the number and 5 stands for the number of hundredths (1/100 or .01 or 10^{-2}). Constructing our place value table we can summarize this as follows:

1	2	3	.	4	5
10^2	10^1	10^0		10^{-1}	10^{-2}
100	10	1		.1	.01
100	20	3		.4	.05

In expanded notation, we write

$123.45 = 1 * 100 + 2 * 10 + 3 * 1 + 4 * (0.1) + 5 * (0.01)$
$ = 1 * 10^2 + 2 * 10^1 + 3 * 10^0 + 4 * 10^{-1} + 5 * 10^{-2}$

We can construct a place value table for the number 892.015 as follows:

8	9	2	.	0	1	5
10^2	10^1	10^0		10^{-1}	10^{-2}	10^{-3}
100	10	1		.1	.01	.001
800	90	2		0	.01	.005

Objective A: Converting Fractional Binary and Hexadecimal to Decimal

How do we convert a binary number that has a fractional part to a decimal number? The same principle we used above applies. The digit immediately to the right of the "binary point" is multiplied by 2, raised to the -1 power, and the next one multiplied by 2, raised to -2 power, etc.

Note: we have used the terminology "binary point" to represent the point separating the whole and the fractional part of our binary number. This is equivalent to using the term decimal point for a decimal number, but since it is a binary number it is called a binary point.

We demonstrate the procedure for converting a fractional binary number to a fractional decimal number by example. We shall convert the binary number 1011.11_2 to its equivalent decimal representation.

Observe that for the place values to the right of the binary point, the exponent is negative. This means that the value is less than 1 and gives us the fractional part of the number. Also the further to the right of the binary point we move, the more negative the exponent becomes. The table below illustrates the first several decimal equivalents to these place values.

Binary Place Value	Decimal Equivalent
$2^{-1} = 1/2$	0.5
$2^{-2} = 1/2^2 = 1/4$	0.25
$2^{-3} = 1/2^3 = 1/8$	0.125
$2^{-4} = 1/2^4 = 1/16$	0.0625
$2^{-5} = 1/2^5 = 1/32$	0.03125
$2^{-6} = 1/2^6 = 1/64$	0.015625
$2^{-7} = 1/2^7 = 1/128$	0.0078125

We begin by constructing a place value table for our fractional binary number

Binary Digits	1	0	1	1	.1	1
Place value	$2^3=8$	$2^2=4$	$2^1=2$	$2^0=1$	$2^{-1}=.5$	$2^{-2}=.25$
Decimal Equivalent	8	0	2	1	.5	.25

and write the expanded form:

$$1011.11_2 = 1*2^3 + 0*2^2 + 1*2^1 + 1*2^0 + 1*2^{-1} + 1*2^{-2}$$
$$= 1*8 + 0*4 + 1*2 + 1*1 + 1*0.5 + 1*0.25$$
$$= 11.75_{10}$$

Example 1 Convert 110.01101_2 to a decimal number

Solution: Construct a place value table for this binary number.

Binary Digits	1	1	0	.	0	1	1	0	1
Place value	2^2=4	$2=2^1$	2^0=1		2^{-1}=.5	2^{-2}=.25	2^{-3}=.125	2^{-4}=.0625	2^{-5}=.03125
Decimal Equivalent	4	2	0		0	.25	.125	0	.03125

To obtain the last row we multiply the numbers in the top row by the value in the second. Then, summing the last row we have:

$$4 + 2 + 0.25 + 0.125 + 0.03125 = 6.40625$$

To convert hexadecimal numbers with a fractional part we follow the same procedure as above, but now we use integer powers of 16. Here is a chart showing the first set of place values:

Hexadecimal Place Value	Decimal Equivalent
16^{-1} = 1/16	0.0625
16^{-2} = $1/16^2$ = 1/256	0.00390625
16^{-3} = $1/16^3$ = 1/4096	0.000244140625

Following the same procedure above we convert the hexadecimal number 9F.B to its decimal equivalent.

Hexadecimal Digits	9	F	.B
Digit value	9	15	11
Place value	16^1=16	16^0=1	16^{-1}=.0625
Decimal Equivalent	$9 \times 16 = 144$	$15 \times 1 = 15$	$11 \times 0.0625 = 0.6875$

$$9F.B_2 = 9*16 + F*1 + B*0.0625$$
$$= 9*16 + 15*1 + 11*0.0625$$
$$= 159.6875_{10}$$

Note: During the calculation we used 15 for F and 11 for B.

Example 2: Convert ABC.E3D$_{16}$ to a decimal number

Solution: Construct a place value table for this binary number.

Hexadecimal Digits	A	B	C	.E	3	D
Digit value	10	11	12	14	3	13
Place value	16^2=256	16^1=16	16^0=1	16^{-1}=.0625	16^{-2}=.00390625	16^{-3}=.000244140625
Decimal Equivalent	2,560	176	12	.9375	.01171875	.00317382125

To obtain the last row we multiply the numbers in the second row by the third Then, summing the last row we have:

2,560 + 176 + 12 + .9375 + .01171875 + .00317382125 = 2,748.95239257

EXERCISES 2.4

Construct a place value table for the given decimal numbers.
1. 429.31
2. 588.1356
3. 2397.15
4. 899.901

Convert each of the fractional binary numbers to a decimal.
5. 1001.0011
6. 1001010.1
7. 11010010.00001

8. 11111.1111
9. 101.101
10. 1 1101.011
11. 1111.1111
12. 11.001

Convert each of the fractional hexadecimal numbers to a decimal.
13. 1AE.11
14. EE.EE
15. ABC.CAB
16. 4D.

Convert each of the fractional octal numbers to a decimal.
17. 324.4
18. 601.2
19. 64.44
20. 27.7

Section 2.5: The Basic Arithmetic of Binary and Hexadecimal Numbers

In this section we take a closer look at how we add and subtract binary and hexadecimal numbers. While addition of positive integers is fairly straightforward, we will see later on that the introduction of negative numbers as well as the operation of subtraction, does introduce some unique challenges that must be overcome.

Objective A: Addition of Unsigned Integers

Let us look at addition in the base 10 or decimal number system. Suppose we want to add 245 and 387, let us review how we do it:

$$\begin{array}{r} 245 \\ +\ 387 \\ \hline 632 \end{array}$$

In base 10, only the numbers 0 through 9 can occupy each place value location. If the sum of two numbers is 10 or more, then we need two place values to represent that sum. In the example above we have, 5 + 7 =12, that means we have one ten and 2 ones. So we put the 2 down (the one's) and carry the 1 (the ten) in the next digit to the left, which is the 10's column.

Next we add the two numbers 4 and 8 to get 12, including the carry 1 to get a sum of 13. This again needs a carry of 1 to the 100's column.

Finally, we add the last two numbers 2 and 3 as well as the carry 1 to give us 6.

The process is exactly the same when adding numbers in the base 2 number system, only now the digit in any one location can only be a 0 or a 1. Thus, if the sum of the two digits adds up to 2 or more, then we have to carry to the next digit (place value).

For example, let us add 1010_2 to 0111_2.

$$\begin{array}{r} {}^{1\ 1} \\ 1010_2 \\ +\ 0111_2 \\ \hline 1\ 0001_2 \end{array}$$

In the first column we add the 1 and the zero and get 1. Moving on to the second column when we add the 1 and the 1 together that gives a value of 2, but since we do not have the symbol 2 in the binary system, we replace it with the binary number for 2 or 10_2. We put down the zero and carry the leading 1 to the next place. In the third column we add together the 1, the zero and the carried 1 to obtain 2 or 10_2 in binary. We put down the 0 and carry the leading 1. We then add the 1 and the 0 and the carried 1 which gives us another 10_2. We put down the 0 and carry the 1, since there is nothing to add the 1 to, we simply write down the 1 for a finished sum of 10001_2.

We can easily check if our addition is correct or not by converting each number to base 10 and then compare the results.

1010_2 is 10 in base 10. 0111_2 is 7 in base 10. The sum of the two numbers 10 and 7 is 17 in base 10. The result, 10001_2 converts to 17 in base 10.

Example 1: Find the sum of

$$1111_2$$
$$+\ 1011_2$$

Solution: In the first column we add the 1's to obtain 2 or 10_2. Write down the 0 and carry the 1. We add the both the 1's in column 2 along with the borrowed 1 to get 3, which in binary is the number 11_2. We write down the 1 and carry the other 1. In the third column we add the 1 and the 0 and the carried 1 to obtain 2 or 10_2. Write down the 0 and carry the 1. In the fourth column we add the two 1's as well as the carried 1 to get 3 or 11_2. Since we do not have any other columns to add we simply write down the 11_2.

$$\overset{111}{1111_2}$$
$$+\ 1011_2$$
$$\overline{1\,1010_2}$$

We can check our result by converting to decimals. The binaries 1111_2 and 1011_2 are equivalent to the decimals 15 and 11, respectively. The sum of 15 and 11 is 26, which is equal to 11010_2.

Let us now look at an example of addition for hexadecimals.

$$589_{16}$$
$$+\ 746_{16}$$

We begin by adding column 1 which is 9+6 = 15, but the hexadecimal symbol for 15 is F. Thus, we write down an F and no carrying is required. We now move to column two. We add 8 and 4 to get 12 or in hexadecimal notation a C. Again, no carrying was required. In the last column we add the 7 and the 5 to get 12, which is again a C in hexadecimal form. Thus the final result is given below.

$$589_{16}$$
$$+\ 746_{16}$$
$$\overline{CCF_{16}}$$

In this example, we did not have to carry to the next place since all the sums were less than 16. If the sum was greater than or equal to 16, then we will have to carry to the next place. We could also convert the hexadecimals to decimals and check our result. The numbers 589_{16} and 746_{16} are equivalent to 1,417 and 1,862 respectively. Their sum is 3,279, which is equivalent to the number CCF_{16}.

Example 2: Compute the given sum
$$ABE_{16}$$
$$+\ BAD_{16}$$

Solution: We begin with the right most column and add the digits E = 14 and D = 13 to obtain 27 which is equivalent to the hexadecimal 1B. Here we write down the B and carry the 1. In column two we add 1, B = 11 and A = 10 to get 22 or written as a hexadecimal it becomes 16_{16}. We write down the 6 and carry the 1. In the last column we add the 1, A = 10 and B = 11, to again obtain 22 or 16_{16}. Here we simply write down the 16 to obtain the result of $166B_{16}$.

$$\overset{1\ \ 1}{A\ B\ E_{16}}$$
$$+\ BAD_{16}$$
$$\overline{16\ 6\ B_{16}}$$

We can also convert to a decimal number to check. Here ABE_{16} and BAD_{16} are equal to 2,750 and 2,989 respectively. Their sum is 5,739, which is equivalent to the hexadecimal number $166B_{16}$.

Objective B: Subtraction of Unsigned Integers

Let us review how we do subtraction in our decimal system. We will consider two types of problems. One that requires borrowing and one that does not.

Subtraction Without Borrowing

First let's subtract 134 from 389. This problem is fairly simple and straightforward, since all the place value digits we are subtracting from are larger than the digits we are subtracting (borrowing is not required.) Thus,

$$\begin{array}{r} 389 \\ - \ 134 \\ \hline 255 \end{array}$$

Similarly, in the binary and hexadecimal system, we follow the same procedure as illustrated in the following examples.

Example 1: Subtract the binary number 1001_2 from 11001_2.

  Solution: This gives us:

$$\begin{array}{r} 11001_2 \\ - \ \ 1001_2 \\ \hline 1\ 0000_2 \end{array}$$

Here we subtract the two 1's in the first column to obtain 0 and then subtract 0 from 0 in column two and three also obtain 0. Then we subtract 1 from 1 in column four to obtain 0 and finally subtract 0 from 1 in column five to obtain 1. The final result is 10000_2.

The problem is similar when working with hexadecimals.

Example 2: Subtract the hexadecimal $A4E_{16}$ from $C8F_{16}$.

  Solution: We obtain:

$$C8F_{16}$$
$$- A4E_{16}$$
$$241_{16}$$

To obtain this result, subtract E = 14 from F = 15 to obtain 1, which is just 1 in hexadecimal notation. Then we move to column two and subtract 4 from 8 to obtain 4, which too is just 4 in hexadecimal notation. Finally we subtract A = 10 from C = 12 to obtain 2. Thus the final answer is 241_{16}. You can also check this by converting to decimal numbers and subtracting. Try this as an exercise.

Subtraction With Borrowing

Now lets consider the more complicated problem of subtraction with borrowing. Again let's review a similar problem in decimal form. Consider subtracting 189 from 234.

This gives the following:

$$\overset{1\ \ 12\ \ 14}{\cancel{2}\cancel{3}\cancel{4}}$$
$$-\ 1\ 8\ 9$$
$$4\ 5$$

Notice we could not subtract 9 from 4 so we had to borrow a 10 from the tens column. Then we could subtract 9 from the resulting 14 to obtain 5. The tens column now has a digit value of 2 (a place value of 20) and we cannot subtract a digit value of 8 (a place value of 80) from the 2. Thus we have to borrow from the 100's place, etc.

We carry out the same process with binary and hexadecimal numbers. Consider the following examples:

Example 3: Subtract the binary number 111_2 from 1000_2.

  Solution: Here we could not subtract the 1 from the 0 in the ones column so we had to borrow from the twos column. The twos column did not have anything to borrow so we had to go down to the fours and eventually the eights. We are essentially borrowing 2 from each column, just as we borrow 10 from each column in base 10. Then we perform the subtraction.

$$\begin{array}{r} {\overset{0}{\cancel{1}}\overset{\overset{1}{\cancel{0}}}{\cancel{0}}\overset{\overset{1}{\cancel{0}}}{\cancel{0}}\overset{2}{\cancel{0}}}_2 \\ -\ 0\ 1\ 1\ 1_2 \\ \hline 0\ 0\ 0\ 1_2 \end{array}$$

The same process is used when subtracting hexadecimals.

Example 4: Find the difference of 388_{16} and $5A6_{16}$.

$$\begin{array}{r} 5A6_{16} \\ -\ 388_{16} \\ \hline \end{array}$$

Solution: We start in column one and try and subtract 8 from 6. The 8 is too large, so we must borrow from the next column. In base 10 we borrow a 10, and in base 2 we borrow a 2, so in base 16 we borrow a 16 from column two. We are borrowing from A = 10 so when we subtract 1 this becomes a 9

We now add the 16 to the 6 to obtain 22 and subtract the 8 which gives us 14 which as a hexadecimal is E. We now move to column two. Here we can subtract the 8 from the 9 to obtain 1. Finally in column 3 we can subtract 3 from 5 to obtain 2. Thus, the resulting hexadecimal difference is $21E_{16}$.

$$\begin{array}{r} \overset{9}{}\overset{1}{}\overset{16+6\ =22}{} \\ 5\ \cancel{A}\ \cancel{6}\ _{16} \\ -\ 3\ 8\ 8\ _{16} \\ \hline 2\ 1\ E\ _{16} \end{array}$$

In the next section we will consider addition of signed numbers where the numbers we are adding can be either positive, negative, or both.

EXERCISES 2.5

Add the following unsigned integers in the bases indicated:
1. $1011\ 0101_2$ $0110\ 1110_2$

2. 1273_8 4251_8

3. $2AD_{16}$ $72F_{16}$

4. $0111\ 0111_2$ $1100\ 0101_2$

5. 5371_8 3157_8

6. ABC_{16} $F4_{16}$

Subtract the following unsigned integers in the bases indicated:
7. $1011\ 0101_2$ $0110\ 1110_2$

8. $1010\ 1010_2$ $0110\ 0110_2$

9. 4321_8 257_8

10. $2A9_{16}$ $1EF_{16}$

11. 3361_{16} BFA_{16}

Add and subtract one from each of the following numbers in its given base.
12. 732

13. 1234

14. 1101_2

15. 10100110_2

16. 235_8

17. 777_8

18. $92A_{16}$

19. $F2F_{16}$

Section 2.6: Computer Arithmetic

One of the fundamental objectives of this book is to understand how computers store and process numbers. We have been working exclusively with unsigned numbers in this chapter. When we showed how to add and subtract positive binary integers we chose problems that always had positive solutions. We did this on purpose. We wanted to present the concept of arithmetic in other base number systems without the extra complications that can arise when performing these operations on a computer with negative results.

We are now ready to take a closer look at how computers process negative numbers. The most elementary operation computers perform on numbers is basic arithmetic. The problem is that unlike humans, computers can only "understand" and process data in bits, (a series of 1's and 0's). There is no concept of a negative number built into the computer. It is something that must be created. We can't write down a negative sign and expect the computer to understand what it means. We have to define it in terms of a bit.

In this section we take a closer look at how negative numbers are represented in a computer. In particular, we look at how signed integers are defined as well as how a fixed and finite number size impacts working with numbers in a computer.

Objective A: Fixed-Bit Representation of Signed Integers and the Two's Complement Method

Computers have a fixed and finite amount of space to store a number. Early computers used a 16-bit storage space for a number. Today's computers use either a 32 or 64-bit storage space for numbers. This is different than what we are used to in mathematics. In mathematics we can write down essentially any number we please. We are not limited to numbers in a certain range or of a fixed length. We discussed this in Section 2.2 when we talked about storing decimal numbers in a computer. We now consider binary numbers and in particular the representation of negative binary numbers in a computer.

Consider a very simple computer that uses 4 bits to represent a number. Obviously, a real 4-bit computer would not be very useful, but for illustration purposes it is easier to present it in this way. With 4 bits there are 16 different patterns that can be used to represent different numbers. This is shown in the table below.

Number	Bit Pattern	Number	Bit Pattern
0	0000	8	1000
1	0001	9	1001
2	0010	10	1010

Number	Bit Pattern	Number	Bit Pattern
3	0011	11	1011
4	0100	12	1100
5	0101	13	1101
6	0110	14	1110
7	0111	15	1111

If we want to represent signed integers we must use one of the bits to represent the sign of the number. The left most bit is used to differentiate between positive and negative integers. If the left most bit is a 0, then the number is positive. If the left most bit is a 1, then the number is negative. Thus by using the left most bit to designate the sign of the number, the largest positive number is now $0111_2 = 7$ not 15.

The remaining positive integers in our 4–bit number are represented just as we would think. The first bit is a 0 indicating the number is positive and the remaining three bits are used to represent the number, as shown in the chart below.

Binary number (base2)	Positive integers
0000	0
0001	+1
0010	+2
0011	+3
0100	+4
0101	+5
0110	+6
0111	+7

To represent negative numbers, the most obvious approach would be to let the first bit be 1, indicating the number is negative, and then let the three remaining bits represent the number, or:

Binary number (base2)	Negative integers
1000	0
1001	−1
1010	−2
1011	−3
1100	−4
1101	−5

Binary number (base2)	Negative integers
1110	–6
1111	–7

This representation could work; however, there is an alternative representation that when implemented on a computer eliminates the need to perform or even define subtraction. Thus, we will not use this representation, but instead introduce an alternative method below.

Two's Complement Method

An alternative approach to representing negative integers is called the **two's complement** method. This method, as we shall see later on, enables us to turn subtraction problems into addition problems. By transforming subtraction into addition we eliminate the need of trying to define the complicated process of borrowing on a computer. This aspect of the two's complement, as we shall show, enables us to create a more efficient method for subtracting numbers in a computer.

The two's complement method is only applied to negative integers. To represent negative integers, we first need to introduce the **one's complement** of a number (not the two's complement, just yet). The one's complement of a binary number is obtained by changing each digit in the binary number to its opposite value. Thus a 1 becomes a 0, and a 0 becomes a 1.

Examples: Find the one's complement of the binary numbers
1. 101
2. 111011
3. 1100111

Solutions:
1. 101 becomes 010
2. 111011 becomes 000100
3. 10111 becomes 01000

To obtain the Two's complement of a number you add 1 to the one's complement of that number.

Examples: Find the two's complement of the binary numbers
4. 101
5. 111011
6. 1100111

Solutions:
      4. 101   one's complement gives 010, and add 1 to obtain 011
      5. 111011 one's complement gives 000100, and add 1 to obtain 000101
      6. 10111 one's complement gives 01000, and add 1 to obtain 01001

Now that we know how to find the two's complement of a negative binary integer, let's look at how we represent a negative decimal integer in binary using the two's complement method. In addition, we'll also work on a computer with a fixed amount of space to represent a number.

Suppose we want to find the two's complement representation of –6 using a 4–bit pattern. To do this we use the following procedure.

1. Ignore the negative sign and convert 6 to binary using 4 bits, i.e. 0110.
2. Take the one's complement of this number, i.e., 1001.
3. Then add 1 at the end.

$$\begin{array}{r} 1001_2 \\ +\ \ 0001_2 \\ \hline 1010_2 \end{array}$$

Therefore, 1010_2 is the two's complement representation of –6 using 4 bits.

Example 7: Write –45 in two's complement representation using an 8–bit pattern.

Solution:
      1. Convert 45 to an 8-bit binary number to get 0010 1101.
      2. Take the one's complement, 1101 0010
      3. Add 1 to it, resulting in 1101 0011.

Therefore, $1101\ 0011_2$ is the two's complement representation of –45 using 8-bit pattern.

The following table summarizes the two's complement representation of signed integers using a 4–bit pattern as well as the corresponding unsigned integers.

Unsigned integer	Binary number (base2)	Signed integer
0	0000	0
1	0001	+1
2	0010	+2
3	0011	+3
4	0100	+4

Unsigned integer	Binary number (base2)	Signed integer
5	0101	+5
6	0110	+6
7	0111	+7
8	1000	–8
9	1001	–7
10	1010	–6
11	1011	–5
12	1100	–4
13	1101	–3
14	1110	–2
15	1111	–1

There are still 16 different numbers that can be represented using 4 bits, but the range is now from –8 to +7 instead of 0 to 15. Since the first bit represents the sign (0 positive, 1 negative) the remaining three bits give the magnitude of the number. By convention the representation for zero is put with the positive numbers and the extra binary pattern is used for the negative numbers.

We see that there is a limited range of numbers that can be represented with any given bit pattern. There is a limit, regardless of how many bits are given. For example, in an eight bit pattern, the range of numbers will be from –128 to +127. This is a total of 256 different signed integers. Compare this with the range of 0 to 255 for the unsigned integers. Note, however, that in both cases the 8–bit pattern can hold up to a maximum of 256 different numbers.

In summary, to write the two's complement representation of a negative integer we:
1. convert the number, ignoring the sign, to binary;
2. take the one's complement;
3. add 1 to the result.

There is an alternative way to find the two's complement of a number. We again ignore the sign and write the equivalent binary number. Now, instead of taking the one's complement and then adding 1, we do the following. We read the binary number we obtained in the first step from right to left until we reach the first 1 value. We keep this part of the number and then take the one's complement of the remaining numbers to the left. Let's illustrate this using some examples.

Example 8: Write –45 in the two's complement representation using an 8–bit pattern and the alternative approach.

Solution: Convert 45 to binary to get 0010 1101. Reading the number from right to left we find the first 1 occurs immediately. We keep this 1 digit and then simply take the opposite of each digit to the left, i.e., 1101 0011. This is the same result we obtained in Example 1 above.

Example 9: Write –92 in the two's complement representation using an 8–bit pattern and the alternative approach.

Solution: First we convert 92 to binary to get 0101 1100. Reading the number from right to left we find the first 1 occurs three numbers from the right. We keep this 100 string of digits and then simply take the opposite of each digit to the left, i.e., 1010 0100.

Objective B: Addition and Subtraction of Signed Integers

In Section 2.5 we showed how to add and subtract unsigned (positive) binary numbers. We could develop a similar procedure for signed integers, but as we said above, borrowing presents a problem that would be especially difficult to implement on a computer.

The two's complement representation provides a unique advantage when the computer has to subtract signed integers. The two's complement approach allows the computer to avoid the whole borrowing process during subtraction. In fact, it allows it to avoid subtraction altogether! To perform subtraction, the computer finds the two's complement of the number it needs to subtract and then adds it to the other number.

For example, let's say we want to perform the following subtraction problem, $15 - 7 = 8$ on a computer that represents numbers with 5-bits.

We start by rewriting the problem as $15 + (-7)$ and perform the addition. Thus, we would have to find the two's complement of –7 and then add the result to the binary representation of 15 (01111_2).

The two's complement of –7 is 11001_2. Thus we have.

$$\begin{array}{r} 01111_2 \\ +\ 11001_2 \\ \hline 101000_2 \end{array}$$

Since this is a 5-bit number, we ignore the leading bit and the answer is 01000_2. This is equivalent to the decimal number 8. Notice that we added the numbers together and there wasn't any complicated borrowing process to perform, as we would have had to do using subtraction.

Using this approach, we never have to worry about borrowing in a subtraction problem. This method also makes computations on a computer much faster as subtraction with borrowing would be far more time consuming than converting and adding. This is precisely why the two's complement method is used on a computer.

Show how the two's complement technique would be used to perform the following problems using 5-bit signed integers.

Example 1: $10 - 14 = ?$

Solution: The first step is to change the subtraction problem into an addition problem by rewriting it as $10 + (-14)$. Using the two's complement method we transform -14 into its two's complement equivalent as follows:
1. write the binary equivalent of 14 as $14_{10} = 01110_2$;
2. takes the two's complement of 01110_2 which is 10010_2.

Next we add this to the 5-bit binary equivalent of $10_{10} = 01010_2$, to obtain:

$$01010_2$$
$$+10010_2$$
$$\overline{11100_2}$$

This number is the two's complement representation of -4.

Example 2: $(-9) - (-2) = ?$

Solution: The first step is to change the subtraction problem into an addition problem by rewriting it as $(-9) + 2$. Using the two's complement method we transform -9 into its two's complement equivalent as follows:
3. Write the binary equivalent of 9 as $9_{10} = 01001_2$.
4. Takes the two's complement of 01001_2 which is 10111_2.

Next we add this to the 5-bit binary equivalent of $2_{10} = 00010_2$, to obtain:

$$10111_2$$
$$+00010_2$$
$$\overline{11001_2}$$

This number is the two's complement representation of −7.

In the examples above we started with a decimal problem, converted it to binary and then performed the arithmetic. In a computer all numbers are stored in binary form. Thus when a computer must subtract two numbers, it only has to apply the two's complement of the number it is subtracting, e.g.,

Example 3: Subtract the 4–bit binary numbers.

$$0111_2$$
$$-\ 0101_2$$

Solution: The first step is to change the subtraction problem into an addition using the two's complement method. We first take the two's complement of the number we are subtracting. The two's complement of 0101_2 is 1011_2. Now we simply add the following two numbers together.

$$0111_2$$
$$+1011_2$$
$$\overline{10010_2}$$

This is a 4–bit number, therefore the first bit is ignored and the result is, 0010_2. The reader should convert this problem to decimals and show that the original problem was $7 - 5 = 2$, and that we obtain the same result.

You must be careful, as the next example shows.

Example 4: Subtract the 4–bit binary numbers.

$$0101_2$$
$$-\ 1101_2$$

Solution: The first step is to change the subtraction problem into an addition using the two's complement method. We first take the two's complement of the number we are subtracting. The two's complement of 1101_2 is 0011_2. Now we add the two numbers together.

$$0101_2$$
$$+\ 0011_2$$
$$\overline{1000_2}$$

This would give us the two's complement representation of –8 in decimal form.

However if we look at the original problem in decimal form it was 5 – (–3) = 5 + 3 = 8 not –8. What happened? This is an example of overflow. Remember that a 4-bit number can only represent integers from –8 to +7. The number 8 is out of the range of the 4-bit number. Thus, there is no solution to this problem.

Objective C: Why the Two's Complement Method Works (Optional)

It may look a little like magic, but there is a logical reason why the two's complement method works for subtraction. To see why, look at the problem above of subtracting the two decimal numbers 189 from 234, but this time we shall introduce a trick that will allow us to avoid borrowing altogether. We simply add and subtract 1000 (equivalent to adding 0) to the problem and then work the remaining problem as follows:

$\quad 234 - 189 + 1000 - 1000$

$= 234 - 189 + 999 + 1 - 1000 \quad$ rewrite the 1000 as 999+1

$= 234 + \mathbf{999} - 189 - 1000 + \mathbf{1} \quad$ rearrange the numbers in slightly different order

$= 234 + \mathbf{810} - 1000 + \mathbf{1} \quad$ subtract 189 from 999 which is 810
(NOTE: It is a lot easier to subtract from 999 than borrowing.)

$= \mathbf{1044} - 1000 + \mathbf{1} \quad$ add 234 and 810 which is 1044

$= \mathbf{44} + \mathbf{1} \quad$ subtract 1000 from 1044 which is 44

$= \mathbf{45} \quad$ This result is the same as what we have above, but in this case we did not have to borrow.

The procedure above involves what is called the ten's complement and the nine's complement. The ten's complements is taking the difference between 1000 and 189, but it is easier to subtract 189 from 999 because it does not involve "borrowing." The difference between 999 and 189 is called the nine's complement.

This is essentially what the two's complement does for binary numbers. Here's how it works:

First you must know how many bits are in the numbers you are subtracting. Let's start with 4–bit numbers. Consider the two numbers 0100 and 0011 where we want to subtract the second number from the first, i.e.,

0100 – 0011

Now lets add and subtract the 5–bit number 10000 (subtracting 0011 from 10000 gives us the two's complement of the second number) Note: if we were dealing with any N–bit number we always add and subtract an (N + 1)-bit number with a 1 for the 1st bit and a zero elsewhere.

This gives us: 0100 – 0011 + 10000 – 10000

Rewriting the first 5–bit number as 1111 + 1, we now have

0100 – 0011 + **1111 + 1** – 10000

We now can subtract the 2nd number from the third without borrowing as follows:

0100 + **1111 – 0011** + 1 – 10000, the number 1111 – 0011 is just the ones' complement of 0011. Adding the additional 1 gives the two's complement.

The expression now becomes: 0100 + 1100 + 1 – 10000

We now add the three first numbers to obtain:

0100 + 1100 + 1 – 10000 = 10001 – 10000

and then subtract the last number from the first to get, 00001, the desired result. In real life, however, we do not have to do the final subtraction. This occurs automatically, since we are using a 4 bit number. This part is simply the computer ignoring the 5th bit.

EXERCISES 2.6

Express each of the decimal numbers in 8–bit, two's complement representation.
1. 91
2. –91
3. 115
4. –37
5. –151
6. 128
7. 65

Express the decimal numbers in 16–bit, two's complement representation.
8. 91
9. –91
10. –151
11. –30,583
12. 40,000
13. –22,559

Find the range of decimal integers that can be expressed with two's complement representation

14. with 32 bits
15. with 64 bits

If possible show how the two's complement technique would be used to solve the following problems using 6-bit signed integers. If not possible state why.

16. 13–26
17. 25+(–30)
18. 8-27
19. (–6)+(–15)
20. 15–(–45)

If possible use the two's complement technique to subtract the 4–bit binary numbers. If not possible state why.

21. $\quad 0011_2 - 0101_2$

22. $\quad 0111_2 - 1001_2$

23. $\quad 0011_2 - 0111_2$

24. $\quad 0100_2 - 0011_2$

If possible use the two's complement technique to subtract the 6–bit binary numbers. If not possible state why.

25. $\quad 010111_2 - 010101_2$

26. $\quad 011101_2 - 010111_2$

27. $\quad 010101_2 - 101101_2$

28. $\quad 011111_2 - 000111_2$

Section 2.7: Applications of Binary and Hexadecimal

Binaries and hexadecimals are used extensively in computers, especially since they are the foundation on which a computer is based. In this section we highlight two of the more common applications beyond the basic arithmetic and memory size from Section 2.1.

Objective A: ASCII and Unicode

The first application of binary and hexadecimal numbers is as a tool for identifying the various characters and symbols printed by a computer. There are two primary standards for identifying characters on a computer. The first is ASCII (American Standard Code for Information Interchange) and the second is a newer more universal standard called Unicode. Both of these standards identify a particular character or symbol by using a specific decimal (base 10), hexadecimal (base 16), octal (base 8) or binary number. For example, the ASCII code numbers used to represent the letter "A", number "3" or symbol "!," are shown below:

Dec	Oct	Hex	Binary	Character
065	101	041	01000001	A
051	063	033	00110011	3
033	041	021	00100001	!

In the ASCII system there are $256 = 2^8$ characters that can be referenced. Typically only half of that list, or 128 characters, are identified. The other 128 characters make up the extended ASCII system. A complete listing of the first 128 characters (0–127) is in Appendix A for your convenience.

Given the hexadecimal numbers, find the associated ASCII character.

Example 1:   2F

   Solution: 2F = 47 in decimal form and this corresponds to the "/" character

Example 2:   1B

   Solution: 1B = 27 in decimal form and corresponds to the "ESC" character

Example 3:   4D

   Solution: 4D = 77 in decimal form and corresponds to the capital letter "M"

As we said above, the ASCII system is only one standard for identifying characters. Another more universal system is **Unicode**. The Unicode system is an international character system and instead of using an 8–bit (2^8, 0–255) number to keep track of characters, the Unicode system uses a 16–bit number to identify all the characters. This allows us to define 2^{16}=65,536 (0–65,535) standard characters.

The reason why there is a movement toward this standard is as follows:

- **Universality**. The list of characters must be large enough to encompass all characters that are likely to be used in general text interchange. This should include all major international, and industrial character sets.

- **Efficiency**. Plain text is simple to search. The software will not have to maintain, state, or look for special escape sequences and character synchronization from any point in a character stream. Character searching is quick and unambiguous.

- **Uniformity**. A fixed character code will allow for efficient sorting, searching, displaying, and editing of text.

- **Eliminate Ambiguity**. Any given 16–bit value always represents the same character.

A consortium was created to come up with and maintain this standard, and for more information, the interested reader can go to: http//www.unicode.org.

Objective B: Memory Identification

Another application of hexadecimal numbers is in identifying memory locations and values on a computer. You have all probably seen a program terminate prematurely and a series of characters appear on the screen. This is a memory dump. Memory dumps can be a programmer's best friend, since they provide information that will enable him/her to determine what may have caused the problem. A memory dump is simply a snapshot of everything in the computer memory at that particular time.

Here is an example of what might appear on a computer screen:

Example: What is stored in the file below and how long is it? What character is represented by the hexadecimal number 00?

File dump — MA17 — 3/8/04
```
0000  00 53 00 74 00 75 00 64 | 00 79 00 20 00 54 00 6F    .S.t.u.d.y. .T.o
0010  00 20 00 50 00 61 00 73 | 00 73 00 20 00 54 00 68    . .P.a.s.s. .T.h
0020  00 65 00 20 00 46 00 69 | 00 6E 00 61 00 6C 00 20    .e. .F.i.n.a.l.
0030  00 45 00 78 00 61 00 6D | 00 21 00 21               .E.x.a.m.!.!
```
File length = 3C

Solution: All the data in this file are given in hexadecimal. Each line contains 16 bytes. The first number in each line is called the "setoff" value which we will ignore. The numbers following this are the 16 bytes of two-digit hexadecimal numbers in memory in 8 byte groupings separated by a "|." Each character is the ASCII representation of that byte, if the byte is printable, otherwise a period is shown if the byte is not printable. Finally, the total length of the file is printed at the end.

Converting the hex numbers to the appropriate ASCII codes, we see that the file reads: "Study To Pass The Final Exam!!." The file contains 3C=60 characters and each letter and space above is separated by the unprintable NULL character represented by 00_{16}.

EXERCISES 2.7

Given the hexadecimal numbers, find the associated ASCII character.
1. 11
2. 43
3. 5E
4. 59
5. 6D
6. 0F
7. 3B
8. 4C

Given the binary numbers, find the associated ASCII character.
9. 0111 0000
10. 0100 0111
11. 0000 1000
12. 1110 0101
13. 0011 1111
14. 1010 0101
15. 1111 0001
16. 0000 0000

COMPUTER ACTIVITY

1. The ASC ("character") command in BASIC enables you to find the decimal value of the ASCII code of any character. For example: ASC ("A") would return the value of 45, since 45 is the decimal ASCII code for the letter A. Write a program that will input any character, and return and print the ASCII code for that character.

Chapter 3

Logic and Computers

Logic plays a vital role in the mechanics and architecture of computers and programming. In this chapter, we will learn the fundamentals of logic required to gain a deeper understanding of computing. We will see how logic enhances our ability to comprehend the topics in this text, as well as form a foundation for further study in the field of computer science and information technology

At this point, we can write mathematical expressions and evaluate them. We can also make decisions about whether mathematical statements are true or not. We cannot, however, compare statements. Logic gives us the tools necessary to make these comparisons and ultimately draw conclusions. Logic provides us with a systematic way to analyze, compare, and finally make decisions regarding statements. Logic is the framework with which we are able to build a decision-making process into a computer and is the fundamental structure behind any computer program.

We begin by introducing the basic terminology and mathematical definitions of logic and show how this applies to computing. Finally, digital circuitry in this era creates the need to understand the " behind the scene" aspects of computing. By the end of this chapter, we will be able to create and simplify a parallel or series electric circuit merely by understanding the mathematics of logic.

After completing this chapter, you should be able to:
- create, identify, and analyze simple and compound statements,
- rewrite sentences and statements symbolically in mathematical terms,
- determine the truth values for logical expressions and mathematical relations that involve various types of connectives including the conjunction, disjunction, conditional, and biconditional,
- construct truth tables logical statements and understand the relationship between a truth table and the binary number system used in computers,
- utilize logical equivalence to compare statements, re-write equivalent statements, and ultimately make decisions based on those statements,
- apply logic to programming in BASIC where a sequence of commands and decisions need to be made, and
- design series and parallel circuits from logical statements and determine the behavior or output for the circuit.

Section 3.1: Logical Statements and Connectives

Objective A: Simple Statements

In this section we focus on the basics of logic. The fundamental building blocks of logic are logical statements.

   Definition: A **statement** is a declarative sentence that is either true or false.

A statement is called simple, or a *simple statement* if there is only a single thought associated with that statement.

Examples:

1. The sentence, "Today is Saturday." is a simple statement.
   This is a statement because today it is either Saturday, hence making the statement true; otherwise, if it is not Saturday, the statement is false. Furthermore, the statement is simple because it is concerned with a single thought.

2. The sentence, " Does everyone love to travel?" is not a statement.
   This sentence is a question that does not have any truth value to it. What is expected is an answer of yes or no. Therefore, this is not a statement. Remember that a statement can always be assigned a truth value.

3. The mathematical expression $2^3 - 3*5$ is not a statement.
   This is an expression that can be evaluated to -7, but this is not a statement, since we cannot assign a truth value to $2^3 - 3*5$.

4. The equation $2^3 - 3*5 = -7$ is a statement.
   The left hand side of the equation is an expression that can be evaluated to -7. It equals the right hand side, therefore, the equation is true. Since it has a truth value, it is statement.

5. The sentence, "Go home!" is not a statement.
   This is command and commands cannot be assigned truth values, therefore, it is not a statement.

Note: Commands and questions are not statements.

In programming, we often use relational operators to make decisions. We will show how we do that later, but for now let's get accustomed to reading statements involving relational operators and determining their truth. You should refer to Chapter 1 where we defined each of them if you need a quick refresher.

Example 5: Determine whether the following are simple statements. If they are, determine their truth value
  a) $(2 < 5)$
  b) $(24 \geq 24)$
  c) $(x^3 = 8)$

Solution:

a) This is a simple statement that is true, since 2 is less than 5.

b) This statement is not a simple statement since there are two thoughts associated here. The statement can be broken down into two simple statements. One statement being 24 is greater than 24, and the other being 24 equals 24. We can say however, that the given statement is true since 24 equals 24 is true. Statements such as this will be discussed shortly.

c) This is not a simple statement because it is not conclusively true or false. Its truth value is dependent only on knowing a particular value of x. As you can see, the statement will be true only if $x = 2$, otherwise it will be false. Statements of this form are called **conditional statements** and, as mentioned above, will be discussed later in the chapter.

Objective B: Connectives and Compound Statements

Compound statements are built from simple statements by using connectives. Since compound statements are themselves statements, they must also have a truth value that can be easily determined. The logical connectives we will be discussing in this chapter include:

- AND
- OR
- NOT
- IF-THEN
- IF AND ONLY IF

We will also refer to logical connectives as *logical operators*. The latter is the computer operation based on the mathematical representation of the connective that will be discussed throughout this chapter.

> ***Definition***: A **compound statement** is statement that consists of at least one simple statement and at least one connective.

We can think of a compound statement as a logical statement with more than one thought associated with it.
The following are examples of compound statements using some connectives listed above

Examples:
1. Today is Saturday AND I will begin to read a new book this morning.

2. Mary is sick AND $2-11=-9$.

3. Next Sunday I will either go fishing on the party boat OR fish from shore.

4. IF it snows tomorrow, THEN I will go skiing.

5. $(7<10)$ OR $(10<3)$.

6. $(51.28>51.275)$ AND $(-74.35<73.035)$.

7. The dog is NOT sleeping.

Compound statements are very common in computing. We often wish to perform a certain task if one or more conditions are met. For example, we can use a computer's logical structure to test predefined conditions and the truth value of this test may execute blocks of code or determine when the program will end.

The following examples will illustrate compound statements consisting of a simple statement and one of the logical operators or connectives listed above. We will cover each of these operators in more detail in subsequent sections; however, it suffices to include these examples now to illustrate how a compound statement is constructed.

The example illustrates how we can write a program that allows the computer to test the truth value of a statement and execute a block or line of code. We will cover IF THEN statements in more detail later in the chapter.

Example 1: Write a program that outputs "hello" whenever a number greater than 10 or less than 0 is entered.

Solution: The two simple statements that make up the compound statement are:
1. The number entered is greater than 10
2. The number entered is less than 0.

The program is of the form:

```
CLS
INPUT " Enter a Real number"; x
If x > 10 THEN PRINT "Hello"
If x < 0 THEN PRINT "Hello"
END
```
You should note that any real number you think of would always be greater than 10, less than 0, or a number between 0 and 10 inclusive.

Statements and logical connectives are the basic building blocks from which more complicated logical arguments and expressions are made. To analyze the resulting arguments and expressions, we re-write the statements in symbolic form. We do this by using a single letter to represent each simple statement. It is customary to represent a simple statement by a lower case letter such as p, q, r, s, and so forth. This will allow us to construct compound statements without having to repeat the same statements or sentences in writing. It will further help in the evaluation of the truth value of the logical expression.

Example 2: Identify the different statements in the sentence: "IF the temperature outside is greater than 75 degrees AND it is sunny, THEN I will wear shorts to work."

Solution: This compound statement consists of three thoughts and makes use of more than one connective. We will use letters p, q, and r to denote the statements as follows:
- p: The temperature outside is greater than 75 degrees.
- q: It is sunny.
- r: I will wear shorts to work.

We can represent the compound statement by the following symbolic expression:

If p and q, then r.

There is common mathematical notation in logic that replaces the words AND and OR. We will encounter these later on in the chapter and show some programs that utilize these logical operators.

EXERCISES 3.1

Determine whether each denotes a statement or not a statement. If it does represent a statement, determine its truth value.

1. Every rational number is an integer.
2. The absolute value of a negative number is a positive number
3. You should go to the Yankee game tonight.
4. The absolute value of every real number is a positive real number
5. $(2 > 3)$
6. $(5-8) = 2(7-3) - 11$
7. $-2 + 7 - 3^2$
8. Are there enough lives left in your video games?
9. $x - 7 = 4$
10. Turn off the Xbox when you are done playing.

Identify the simple statements from the given statements below.

11. $5 = 7$ or $(2 \geq -2)$.
12. $("Dad" \leq "Daddy")$ and $("Mom" < "Mommy")$
13. If an integer is odd, then the square of that integer is also an odd integer.
14. If $x = 7$ then $3x - 1 = 20$

Identify the statements as simple or compound.

15. $5 + 3 = 8$
16. $(25 < -25)$ or $(23 > 12)$
17. New York is the capital of Italy or $7 > -20$
18. $(-15 < -25)$
19. Labor Day does not fall on the month of August.

Write each of the following in English sentences, if p= I will get a 90 on this exam, q = I will get an A in the course, and r = I will take another math course.

20. p and q
21. If p then r
22. If p or q, then r.
23. If r then p.

Write the following sentences as logical expressions by first identifying and labeling the statement by a lower case letter.

24. If I drive to school, then I will need to have a driver's license.
25. If you are drinking and driving, then you are breaking the law.
26. If your semester g.p.a. is at least 3.5 and you carry at least a 12 credit load, then you are eligible for the Dean's List

Determine the simple statements and the logical connective for the following compound statements.

27. $(10 > 15)$ AND $(7 < 20)$
28. $(5+3) > 7$ OR $(12+4) < 10$
29. $(-5 < 9)$ and $(2 \neq 11)$
30. $[(2 \geq -15)$ or $(7 < 15)]$ and $(16 \neq 32)$

Section 3.2: Conjunction and Disjunction

In section 3.1 we introduced compound statements. Compound statements consist of simple statements connected with words such as AND and OR. This includes sentences such as, " I will study mathematics and information technology," or, " I will play soccer or go fishing." In this section, we will discuss the AND and the OR connectives in detail. We shall also show when the compound statements associated with these connectives are true or false.

Objective A: Disjunction and the Logical Operator OR

The sentence, "I will play soccer or go fishing," has two thoughts connected with the word OR. In logic, we express the logical operator OR as the disjunction.

Definition: The **disjunction** of two simple statement p, q, denoted by $p \vee q$, read as " p OR q" is the compound statement that is true when p is true, q is true, or both are true.

An alternative way to view this is to say the disjunction of two statements is true except when both statements are false.

When we are interested in the truth value of a compound statement, we must take into account all possible truth values that can be assigned to each of the statements. The tree diagram below shows the different cases that are possible for a compound statement that consists of two simple statements:

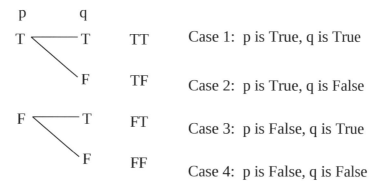

In addition to tree diagrams, we can also use a truth table to outline the various possible cases. A truth table also contains a column for the truth value of the compound statement. The table below outlines all possibilities along with the final answer column for the disjunction $p \vee q$.

p	q	p∨q
T	T	T
T	F	T
F	T	T
F	F	F

Example 1: *Determine the truth value of the compound statement:*
  Jersey City is the capital of New Jersey or John F. Kennedy was a President of the United States.

Solution: Trenton is the capital of New Jersey, not Jersey City. However, John F. Kennedy was a president of the United States. So one of the two statements is true; therefore, the compound statement $p \vee q$ is true.

Example 2: *Determine the truth value of the compound statement:*
  Rome is in France or 5 + 2 = 10.

Solution: Both of the simple statements that form the disjunction $p \vee q$ are false, therefore the compound statement is false.

Example 3: *If C = 5, and D = 7, Determine the truth value of*
  a) $(C < 4) \vee (D \geq 10)$
  b) $(C \geq 5) \vee (D = 8)$
  c) $(C \leq D) \vee (2D > C)$

Solutions:

a)
$$(C < 4) \vee (D \geq 10)$$
$$(5 < 4) \vee (7 \geq 10)$$
$$F \quad \vee \quad F$$
$$F$$

b)
$$(C \geq 5) \vee (D = 8)$$
$$(5 \geq 5) \vee (7 = 8)$$
$$T \quad \vee \quad F$$
$$T$$

c)
$$(C \leq D) \vee (2D > C)$$
$$(5 \leq 7) \vee (2 \times 7 > 5)$$
$$(5 \leq 7) \vee (14 > 7)$$
$$\text{T} \quad \vee \quad \text{T}$$
$$\text{T}$$

Objective B: Conjunction and the Logical AND

We will now discuss compound statements with the AND connective. We will refer to the connective AND as a logical connective or an operator interchangeably.

Definition: The **conjunction** of two simple statements, p, q, denoted by $p \wedge q$, and read as " p AND q", is the compound statement that is true only when both p, q are true and false otherwise.

The following truth table illustrates the possible truth values for the conjunction depending on what the truth values of the simple statements are.

p	q	$p \wedge q$
T	T	T
T	F	F
F	T	F
F	F	F

Example 1: Determine the truth value of the compound statement:
The Empire State Building is in New York City AND 2 + 2 = 4.

Solution: Both simple statements are true, therefore we say the conjunction is true.

Example 2: Determine the truth value of the compound statement:
There are twelve months in a year AND Princeton University is in Connecticut.

Solution: The first statement is true since there are twelve months in a year but Princeton University is not in Connecticut, it is in New Jersey; therefore, the conjunction is false.

Example 3: Suppose p, q, r represent the following statements.

    p: $15 < 12$
    q: $11 \geq 10$
    r: $22 + 18 = 52 - 12$

Determine the truth value associated with each logical statement below:
a) $p \wedge q$
b) $p \vee q$
c) $(p \wedge q) \vee r$

Solution: We first need to determine the truth value for each of the simple statements above.

    p is False since 15 is not less than 12.
    q is True since 11 is greater than or equal to 10.
    r is True since 40 = 40.

These truth values will help us to determine the truth value for the compound statements in parts a)., b)., and c). above.

The preceding can be summarized as follows: p = F, q = T, r = T

a)
    $(p \wedge q)$
    $(F \wedge T)$
      F

b)
    $(p \vee q)$
    $(F \vee T)$
      T

c)
    $(p \wedge q) \vee r$
    $(F \wedge T) \vee T$
      F $\vee$ T
        T

Example 4: Determine whether the statements are true or false.
a) $(2 < -3)$ and $(15 < 4)$
b) $(2 < 3)$ or $(25 < 4)$

Solution:

a) To determine the truth value of the statement $(2 < -3)$ and $(15 < 4)$, we examine the truth values of the logical operator AND. We will proceed as before.

$$(2 < -3) \text{ and } (15 < 4)$$
$$\text{F AND F}$$
$$\text{F} \wedge \text{F}$$
$$\text{F}$$

The statement is false.

b) To determine the truth value of the statement $(2 < 3)$ or $(15 < 4)$, we examine the truth values of the logical operator OR. We will proceed as before.

$$(2 < 3) \text{ OR } (15 < 4)$$
$$\text{T} \vee \text{F}$$
$$\text{T}$$

The statement is true.

Logical operators such as AND and OR are regularly used when a search is conducted using an Internet search engine. Suppose you want to do a search on the Internet for mathematics AND computing. You might use any search engine you like such as Google, Yahoo, Hotbot, etc. and type the key word of interest and observe the number of hits (websites). Those websites listed contain both of the words mentioned. If we change the search field to read mathematics OR computing, the resulting websites will include either one or both of these keywords. You should certainly expect to get a lot more hits when you use the OR operator in the search field.

The following example illustrates the above, but you should note that the number of hits changes daily as websites are updated, deleted, or added regularly.

For example: A search was performed on Google.com for (mathematics) AND (computing) that resulted in approximately 2,810,000 hits and a search for (mathematics) OR (computing) resulted in approximately 8,970,000 hits. You should try variations of the keywords and logical operators and observe the number of hits you get for each search.

Objective C: Programming Applications

We introduced some programming concepts in the previous chapters. In this section, we will elaborate and use the logical operators discussed so far to determine the truth value of compound statements. When mathematical relations such as equalities and inequalities along with connectives such as AND and OR are used in BASIC, the result is always a truth value. BASIC indicates a true statement with a 1 and a false statement with a 0. That is, the program will output a 1 if the statement is true and a 0 if the statement is false.

In some versions of BASIC, a –1 rather than a 1 may represent a true statement. In this text, a true statement will be represented as a 1 when referring to the output of a program.

The way of representing mathematical relations is different in BASIC than it is with mathematical notation. The table below will help with the correct syntax when programming.

Mathematical Notation	Computer Syntax	From left to right read as:
$<$	<	Less than
$\leq$	<=	Less than or equal to
$>$	>	Greater than
$\geq$	>=	Greater than or equal to
$=$	=	Equal to
$\neq$	<>	Not equal to

The following examples outline how we can use the computer to determine the truth values assigned to mathematical relations containing logical operators. Enter and execute each of the programs to verify the results.

Example 1:
```
CLS
PRINT (2<=3)
END
```

Solution: Since this is a true statement, the output will be a 1.

Example 2:
```
CLS
PRINT (2>3) AND (15<>4)
END
```

Solution: Since this is a false statement, the output will be a 0.

Example 3:
```
CLS
PRINT (2<3) OR  (4>=25)
END
```

Solution: Since this is a true statement, the output will be a 1.

How do we compare character strings? When the statements contain characters, the ASCII binary representation of the value is determined and used to compare. We discussed the ASCII system in section 2.8 of this text and you can also reference the back of this text for the ASCII table of values.

Example 4:
```
CLS
PRINT ("A" < "B")
END
```

Solution: The result of this is true since the ASCII binary value of A (0100001) is smaller than the ASCII binary value of B (01000010).

Example 5:
```
CLS
PRINT ("a" < "A")
END
```

Solution: This program will output a 0 since it is a false statement. The ASCII binary representation of "a" (01100001) is greater than the ASCII representation of A (01000001).

The necessity of the quotation marks arises above since we are relating one character to another character. The inequality relation then uses the binary representation of these characters to determine the truth value of the relation. The comparison is done on a bit by bit basis.

Example 6 will illustrate what happens when one side of the relation contains more characters than the other side.

Example 6:
```
CLS
PRINT ("ab" < "ac")
END
```

Solution: This program will output a 1 since it is true.
The reason for this is that BASIC compares the ASCII characters in this relation bit by bit. If the first characters are the same, the second characters will decide the order. In this case, since the first character "a" is the same, the second character decides order. Since the ASCII binary value of "b" is less than "c", the relation `"ab" < "ac"` is true. The resulting output will be 1.

Generally, when determining character relations, the first different character decides the order.

EXERCISES 3.2

Write the following statements symbolically using letters and the connectives $\vee$ and $\wedge$.

1. The computer is on or the printer is broken.
2. Today is a holiday and classes are cancelled.
3. Mary is the class president and she likes classical music.
4. Sue and Dan are coming to dinner or we are going to watch television.
5. You study for the test or you will fail the course.
6. There is a storm and the electricity went out.

Given p is true, q is false, and r is false, find the truth values of the statements below:

7. $(p \wedge q)$
8. $(q \vee r)$
9. $(p \wedge r) \vee q$

Let p: Information Technology students take MA17 and q: MA17 is a corequisite of Visual Basic programming. Write each of the following symbolic expression in sentences.

10. $(p \wedge q)$
11. $(p \vee q)$

Given $p: -3 \leq -5$, $q: (17.3 < 17.03)$, *and* $r: x^2 > 0$, *where x is a real number, determine the truth value of each of the following:*

12. $(p \wedge r) \vee q$

13. $(q \vee r) \vee p$

COMPUTER ACTIVITY

1. Write a program that determines the truth value for the following statements by first assigning the values A = 5, B = 2, and C = 10.

a) $(A \geq B) \vee (C < A)$
b) $(A < B^2 + 2) \wedge (C > B)$
c) $((C - B) < 5) \vee (A = (B - C))$

2. Modify the program above to ask the user for the values of A, B, and C using INPUT statements.

Section 3.3: Negation of Statements

In this section we will discuss the negation or "NOT" operator. We also discuss quantifiers (such as all, none, and some) and the role they play in logic and computing. Finally, the negation of quantifiers will be presented.

Objective A: Negation of Statements and the Logical Operator NOT

In the previous section we introduced the connectives AND and OR and we used the connectives to create new statements. Another way to create a new statement from a given statement is to use the negation or NOT operator.

> *Definition:* The **negation** of a statement p, denoted by ~p, is False if p is true and true if p is false.

Examples: Write the negation of the statements below.

1. p: I will read my mathematics text after I eat.
2. p: The grass is green.
3. p: $(x \geq 3)$; $x \in R$
4. p: Tetris is a video game.

Solutions:

1. The negation of statement p, denoted by ~p can be read as, " It is not true that I will read my mathematics text after I eat," or can be restated as, "I am not going to read my mathematics text after I eat." There are more ways to restate ~p in English. Try to come up with some on your own.

2. The negation of this statement is: The grass is not green.

3. The statement p indicates that any real number greater than or equal to three would make p true. The negation of p would therefore be any real number less than three. Mathematically we can write ~p as $(x < 3)$; $x \in R$.

4. The negation of this statement is: Tetris is not a video game. Or we can say: It is false that Tetris is a video game.

In computing, the NOT operator is the operator we use for negation of statements. Consider the following example:

Example 5: Determine the output of the program
```
CLS
PRINT NOT (5 = 3)
END
```

Solution: The result will be 1 since the statement in the program is true. You should observe that the statement (5 = 3) is false, and therefore NOT (5 = 3) is true.

Example 6: Determine the output of the program
```
CLS
PRINT NOT (2 < 3)
END
```

Solution: The result will be 0 since the statement in the program is False. You should observe that the statement (2 < 3) is True, and therefore NOT (2<3) is False.

Objective B: Quantifiers and Negation of Quantified Statements

Thus far, we have introduced the concept of statements and how to negate statements. We now discuss statements that use words such as "some," "all," and "there exists," as well as the negation of those statements. Just as we have done before, we will also be able to determine the truth value of those statements.

The Existential Quantifier means, "There exists," Other terms commonly used include "some" and "at least one."

For example, the statement, "Some people are friendly," illustrates the existential quantifier "Some." The word "some" means there is at least one person that is friendly. Furthermore, the fact that there is at least one indicates that one or more or all people are friendly.

There is a difference in English as well as mathematical logic when the word "some" is replaced with the word "all" in the above example.

Let's consider the statement, "All people are friendly." This means that there are no unfriendly people. This is a powerful statement and commonly used in mathematics; however, caution should be used when used as part of an English statement.
You might hesitate in saying that, "All students go dancing on Friday night," is true but may feel more confident in the statement, "Some students go dancing on Friday night," being true.

The <u>Universal Quantifier</u> that means "For All" or "For Every." For example, the statement, "All integers are real numbers," utilizes this quantifier.

Caution must be applied when negating quantified statements. The next few examples will illustrate the negation of quantified statements.

Consider the statement p: All integers are rational numbers. We can then state ~p in one of the following forms:

      ~p: Not all integers are rational.
           Or
      ~p: Some integers are not rational.
           Or
      ~p: There exists at least one integer that is not rational.

It should be clear that p is a true statement since every integer can be expressed as a rational number. You can see this since the integer 5 can be written as $\frac{5}{1}$; a rational number. Since p is a true statement, ~p must be false.

Consider the statement p: All dogs can fly.

We would like to see what the negation of the statement p is. Initially you might make an error and state that ~p says, "No dogs can fly." Suppose statement p above is true. The existence of only one dog would make p a false statement. Therefore we say ~p: There exists a dog that cannot fly.

Another similar type of example is the statement q: All dogs chase cats. Again, an incorrect interpretation of ~q is, "No dogs chase cats." Its negation is merely found by stating the existence of at least one dog (one or two, or all dogs will also do) to fit the criteria that it does not chase cats. With all this said, ~p can be stated in any of the following ways:

      ~p: Not all dogs can fly, or
      ~p: There exists a dog that does not fly, or
      ~p: Some dogs cannot fly, or
      ~p: At least one dog cannot fly.

Moreover, if we want to negate the statement r: Some dogs could fly. It would be ~r: All dogs cannot fly.

Examples: State the negation of the statements below.
1. p: All integers are positive.

2. q: Some real numbers are irrational

Solutions:
1. The negation of p is the statement ~p: There exists some (one or more) integer that is not positive.

2. The negation of q is the statement ~q: No real number is irrational.

The following table will help in negating statements with quantifiers.

Statement: p	Negation of Statement: ~p
All do	Not all do
	Some do not
	There exists some that do not
Some do	None do
There exists some	None exists
None do	At least one does
Some do not	All do

Note: We have introduced only the fundamentals of quantifiers in this section. A complete discussion on quantifiers is beyond the scope of this text and are discussed in detail in textbooks such as logic and set theory.

EXERCISES 3.3

Write the negation of each of the statements and determine the truth value of the negation statement.

1. p: Some computers are outdated.
2. q: $-5 \geq 12$
3. r: $24 > 15$

4. s: All students in this class are Information Technology majors.
5. t: At least one integer is the solution to the equation $x^2 > -8$.
6. w: Selden is the capital of New York State.
7. x: The Yankees are a football team.
8. y: 6 < 10
9. z: All natural numbers are positive.
10. m: Some real numbers are not rational numbers.

Let p: Mary takes discrete mathematics and q: Mary is a computer science major, write each of the following symbolic expressions in sentences.

11. $p \wedge q$
12. $p \vee q$
13. $\sim p \wedge \sim q$
14. $p \vee \sim q$
15. $\sim (p \wedge q)$

Let p: $x \leq 5$ and q: $y > 0$. Determine the truth value of each of the following for x = 3 and y = 8.

16. $p \wedge q$
17. $p \vee q$
18. $p \vee \sim q$
19. $\sim (p \wedge q)$
20. $\sim (p \vee q)$

COMPUTER ACTIVITY

What will be the output when each of the following codes is run? You should expect a 0 if it is False and 1 if it is True.

1.
```
CLS
PRINT NOT (2<3)
END
```

2.
```
CLS
    Print NOT (4 < 3) AND NOT (7>2)
    END
```

3.
```
CLS
    INPUT X
    PRINT (X > 0) AND NOT (X < = 5)
    END
```

4. This exercise requires you to use the Internet. Use any search engine of your choice and record the number of results that has the keyword(s) that you enter for each case below.
    i.   mathematics
    ii. mathematics OR discrete
    iii. mathematics AND discrete
    iv. mathematics AND calculus
    v.   mathematics AND (discrete OR calculus)
    vi. NOT mathematics

Write a program that determines the truth value for the following statements assuming A = 15, B = 12, and C = −4.

1. $(A \geq B) \vee \sim (C < A)$
2. $\sim (A < B^2 + 2) \wedge (C > B)$
3. $((C - B) < 5) \vee \sim (A = (B - C))$

Write a program that asks the user for the values of A, B, and C using INPUT statements.

Section 3.4: Tree Diagrams and Truth Tables

In this section, we will discuss how to construct truth tables that can be used to evaluate and analyze more complicated statements.

A systematic approach to constructing truth tables is very useful in the analysis and determination of truth values of complex logical expressions. The student should practice reproducing the tables exactly as they appear in the text. Doing so will ensure that each possible case is taken into account. The number of possible cases in a truth table is dependent on the number of distinct simple statements in the logical expression.
We shall demonstrate the number of possible cases using a tree diagram and show how we transfer the information from a tree diagram to a truth table.
A truth table contains all possible truth values a statement can have. The table below illustrates the possible truth values for a simple statement p.

We know that a simple statement is either true or false as shown in the truth table below.

p
T
F

To illustrate the truth table for $\sim p$, we need to include another column along with its corresponding truth values. The table below is listed for reference.

p	$\sim p$
T	F
F	T

We have seen earlier that the use of a tree diagram is a nice aid in listing all possibilities of truth values whenever we have a logical expression with multiple statements (multiple letters). The tree diagram below begins with the possible truth values of p in the first column, then the possible truth values of q given each possible outcome of p in the second column, and so forth for r and s in the third and fourth columns, respectively.

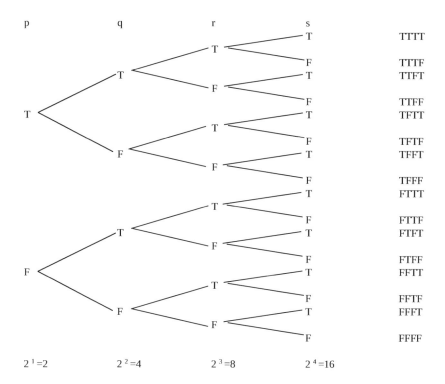

Also, note that the bottom of the tree diagram indicates the number of all possible outcomes of truth values depending on the number of simple statements. In this case, there are four simple statements and hence, there are 2^4 or 16 possible outcomes. If there are five simple statements in a logical expression, then there will be $2^5 = 32$ possible outcomes.

Truth Tables of Compound Statements

Every simple statement is either true or false. We have already shown that the truth tables for conjunction and disjunction have four possible cases for two simple statements. Also, for the negation of a simple statement, there are only two cases. Now, we extend this to more than two simple statements.

Consider the now familiar statement $p \vee q$. This statement has two simple statements p, q. Since there are two simple statements, the truth table will have 2^2 or 4 rows. We have already said earlier that the only time that a disjunction will be false is when both p and q are false.

p	q	p ∨ q
T	T	T
T	F	T
F	T	T
F	F	F

Notice that all possibilities of p and q are listed. Also note that since there are 4 rows in the table, the first column lists the first half of the truth value possibilities as T and the other half as F. The second column entries alternate between T and F beginning with T. The last column indicates the truth value when the logical operator is applied to each case. Let us summarize the truth tables for disjunction, conjunction, and negation. These rules will be used when we construct truth tables of more complicated logical expressions.

p	q	disjunction p ∨ q	conjunction p ∧ q	negation ~p
T	T	T	T	F
T	F	T	F	F
F	T	T	F	T
F	F	F	F	T

The truth table for a logical statement with three distinct simple statements such as $(p \vee q) \vee \sim r$ will have 2^3 or 8 rows. The standard truth table with three distinct simple statements (three letters) can be constructed as follows:

Step 1: The first column will have the first half (or 4 at a time) rows to be T and the other half will be F.

Step 2: The second column will begin with T taken two at a time and alternating with F taken two at a time.

Step 3: The final column will alternate between T and F beginning with T.

The truth table for $(p \vee q) \vee \sim r$ will be:

Step 1:

p	q	r	(p ∨ q) ∨ (~r)
T			
T			
T			
T			

p	q	r	(p ∨ q) ∨ (~ r)
F			
F			
F			
F			

Step 2:

p	q	r	(p ∨ q) ∨ (~ r)
T	**T**		
T	**T**		
T	**F**		
T	**F**		
F	**T**		
F	**T**		
F	**F**		
F	**F**		

Step 3:

p	q	r	(p ∨ q) ∨ ~ r
T	T	**T**	
T	T	**F**	
T	F	**T**	
T	F	**F**	
F	T	**T**	
F	T	**F**	
F	F	**T**	
F	F	**F**	

You should notice the pattern of developing the truth table for consistency.

Finally, we complete the truth values for the logical expression $(p \vee q) \vee \sim r$. To do this, we first fill in the truth values of $(p \vee q)$, and then $\sim r$. The final column will be the disjunction of these two statements.

p	q	r	(p ∨ q)	~ r	(p ∨ q) ∨ ~ r
T	T	T	T	F	T
T	T	F	T	T	T
T	F	T	T	F	T
T	F	F	T	T	T
F	T	T	T	F	T
F	T	F	T	T	T
F	F	T	F	F	F
F	F	F	F	T	T

Example 1: Construct and complete the truth table for the logical statement
$(\sim p \vee q)$.

Solution: Since there are two distinct simple statements (two letters) the truth table for this logical statement will have 2^2 or 4 rows. The table is completed below.

p	q	$\sim p$	$\sim p \vee q$
T	T	F	**T**
T	F	F	**F**
F	T	T	**T**
F	F	T	**T**

Example 2: Construct and complete the truth table for the logical statement
$(\sim p \vee q) \wedge [(r \vee s) \vee p]$.

Solution: Since there are four distinct simple statements (four letters) the truth table or this logical statement will have 2^4 or 16 rows. The table is completed below.

p	q	r	s	$\sim p$	$\sim p \vee q$	$(r \vee s)$	$(r \vee s) \vee p$	$(\sim p \vee q) \wedge [(r \vee s) \vee p]$
T	T	T	T	F	T	T	T	**T**
T	T	T	F	F	T	T	T	**T**
T	T	F	T	F	T	T	T	**T**
T	T	F	F	F	T	F	T	**T**
T	F	T	T	F	F	T	T	**F**
T	F	T	F	F	F	T	T	**F**
T	F	F	T	F	F	T	T	**F**
T	F	F	F	F	F	F	T	**F**
F	T	T	T	T	T	T	T	**T**
F	T	T	F	T	T	T	T	**T**
F	T	F	T	T	T	T	T	**T**
F	T	F	F	T	T	F	F	**F**
F	F	T	T	T	T	T	T	**T**
F	F	T	F	T	T	T	T	**T**
F	F	F	T	T	T	T	T	**T**
F	F	F	F	T	T	F	F	**F**

The study of computers in requires knowledge of the binary number system that we presented in Chapter 2. As you recall, this system contains the digits 0 and 1 only. Since a statement p takes on one of two truth values and a switch can be either on or off, a truth table can be used to also analyze current flow of an electric circuit based on the position of each of the switches. An electric circuit then can be written using a combination of OR, AND, and NOT much like a logical expression. We can follow the rules of constructing a truth table to figure out the results of the output of any electric circuit. The circuits will be discussed more in detail in section 3.7.

In the following table, p, q, r represent switches. Just like the truth table for a logical expression, there are 8 possibilities. Instead of T, we use 1, and instead of F, we use 0. We follow the rules for disjunction if the switches are connected with the word OR. We follow the rules for conjunction if the switches are connected by the word AND.

p	q	r	Rule for Circuitry (Output Result)
1	1	1	
1	1	0	
1	0	1	
1	0	0	
0	1	1	
0	1	0	
0	0	1	
0	0	0	

The next example lists a possible scenario of a student claiming to be eligible to graduate with honors. To be eligible for graduation with honors, a student must take at least two honors courses and maintain a g.p.a. of 3.2 in those courses as well as a 3.0 overall g.p.a.

Example 3: Use a truth table to list all possible cases for when a student can graduate with honors.

Solution: There are three simple statements we need to consider.
p: Two honors courses were taken.
q: The student g.p.a. in the honors courses is at least a 3.2
r: The student has at least a 3.0 overall g.p.a.

We will construct the truth table indicating a 1 when each condition is satisfied and a 0 if the condition has not been satisfied. Finally, we will determine the truth values of the statement $p \wedge q \wedge r$. The result of a 1 in the final column will indicate that the student will graduate with honors.

p	q	r	p∧q	(p∧q)∧r
1	1	1	1	**1**
1	1	0	1	**0**
1	0	1	0	**0**
1	0	0	0	**0**
0	1	1	0	**0**
0	1	0	0	**0**
0	0	1	0	**0**
0	0	0	0	**0**

Notice that the 1 appears only in the first column. This indicates the only possible way the student will graduate with honors is when all three statements are satisfied, hence, true. This is of course what we expected to see.

We can also list all possible cases when a statement is true when the example is not as clear as the previous example was.

Example 4: Using a truth table, list all cases that $(p \vee q) \wedge r$ will result in a true statement; given that p: Jack is Studying,
$q:(x>5)$ and
r = I love Mathematics.

Solution: We will construct the Truth table for $(p \vee q) \wedge r$. The last column will be the results of $(p \vee q) \wedge r$.

p	q	r	p∨q	(p∨q)∧r
1	1	1	1	**1**
1	1	0	1	**0**
1	0	1	1	**1**
1	0	0	1	**0**
0	1	1	1	**1**
0	1	0	1	**0**
0	0	1	0	**0**
0	0	0	0	**0**

The statement is True in cases 1, 3, 5 only and False otherwise.

EXERCISES 3.4

Construct a truth table for each of the following logical statements:

1. $p \wedge \sim q$
2. $(\sim p \vee q) \vee p$
3. $(p \wedge q) \vee (\sim p \vee q)$
4. $p \vee (\sim q \wedge r)$
5. $\sim (p \vee q)$
6. $\sim p \vee (q \wedge \sim r)$
7. $(p \wedge q) \vee (\sim r \wedge s)$

If p and q are true and r is false, determine the truth value of each of the following statements:

8. $\sim p \vee q$
9. $q \wedge \sim r$
10. $(\sim q \wedge r) \vee (p \wedge r)$
11. $p \vee [q \wedge (\sim r \vee \sim p)]$
12. $\sim (p \wedge q) \vee (\sim p \vee r)$

Let p: 4 < 7, q: 8 > 10, and r: 5 > 0. Determine the truth value of each of the following statements:

13. $p \wedge q$
14. $q \vee r$
15. It is false that (8 > 10 and 5 > 0).
16. 5 > 0 or (4 < 7 and 8 > 10).

Let p: Peter is an information technology major and q: Peter is taking MA17, express each of the following symbolic expressions in sentences.

17. $p \vee q$
18. $\sim p$
19. $p \wedge \sim q$
20. $\sim p \wedge q$
21. $\sim p \vee \sim q$

Section 3.5: Conditional and Biconditional

We have discussed the connectives OR, AND, and NOT in the previous sections. In this section we'll learn about two other connectives, the conditional and the biconditional. The conditional is of the form IF *statement 1*, THEN *statement 2* while the biconditional is of the form *statement 1* IF AND ONLY IF *statement 2*. We will concentrate more on the conditional because this is one of the most important concepts in logic as well as computing.

Conditional statements are also commonly found in postulates, theorems, and axioms in mathematics. Below are some of the postulates for equality and inequality.

Postulates of Equality:
If $a = b$ then $b = a$
If $a = b$ and b=c then a=c

Postulates of inequality
If $a < b$ then $a + c < b + c$
If $a > b$ then $a + c > b + c$
If $a < b$ then $a*c < b*c$; c>0
If $a > b$ then $a*c > b*c$; c>0

Objective A: The Conditional Statement

Definition: The **conditional** of two statements p, q, denoted by $p \rightarrow q$, read as, " If p then q, ", or, " p implies q," is the statement that is false only when p is true and q is false, and is true otherwise.

The following truth table outlines the definition.

p	q	$p \rightarrow q$
T	T	T
T	F	F
F	T	T
F	F	T

To help in the understanding of this definition, let's consider the following example:
        p : It rains today
        q : I will go to the movies
        Where $p \rightarrow q$: If it rains today, then I will go to the movies.

We will look at this statement as a promise made to ourselves. To help us understand the definition of the conditional statement, we will look at each of the possible cases. A broken promise will result in a false statement and a promise that has not been broken will result in a true statement.

Case 1: p is true and q is true.
> That means that it is raining today and I did go to the movies. My statement $p \rightarrow q$ is true as I did what I said I was going to do. I kept my promise.

Case 2: p is true and q is false.
> Since it is raining is True and I did not go to the movies, I did not keep my promise. The statement $p \rightarrow q$ is then a false statement.

Case 3: p is false and q is true.
> It did not rain today but I did go the movies is a true statement since my promise said nothing about what I would do if it did not rain. My promise was not broken and the statement $p \rightarrow q$ is a true statement.

Case 4: p is false and q is false.
> My statement $p \rightarrow q$ is true.
> It did not rain, and I did not go to the movies. Again, my promise was not broken since I said nothing about what I would do if it were not raining. Therefore, my statement $p \rightarrow q$ is indeed true.

Adding to the truth tables we constructed above for the connectives OR, AND, and OR, we have the following table:

p	q	disjunction $p \vee q$	conjunction $p \wedge q$	negation $\sim p$	conditional $p \rightarrow q$
T	T	T	T	F	T
T	F	T	F	F	F
F	T	T	F	T	T
F	F	F	F	T	T

We can construct truth tables for logical expressions that includes the conditional.

Example 1: Construct and complete the truth table for the logical statement
$(p \vee q) \rightarrow p$.

> *Solution*: Since there are two distinct simple statements (two letters) the truth table for this logical statement will have 2^2 or 4 rows. The table is completed below.

p	q	p ∨ q	(p ∨ q) → p
T	T	T	T
T	F	T	T
F	T	T	F
F	F	F	T

The last column will always have the statement that we wish to find the truth values of. In this case, the final column is a result of $(\text{column3}) \to (\text{column1})$.

Example 2: Construct and complete the truth table for the logical statement $(p \to q) \wedge (p \to r)$.

Solution: Since there are three distinct simple statements (three letters) the truth table for this logical statement will have 2^3 or 8 rows. We need to construct a column for $p \to q$, a column for $p \to r$, and finally a column for $(p \to q) \wedge (p \to r)$ and indicate their corresponding truth values for each possible case. The table is completed below.

p	q	r	p → q	p → r	(p → q) ∧ (p → r)
T	T	T	T	T	T
T	T	F	T	F	F
T	F	T	F	T	F
T	F	F	F	F	F
F	T	T	T	T	T
F	T	F	T	T	T
F	F	T	T	T	T
F	F	F	T	T	T

Example 3: Construct and complete the truth table for the logical statement $(\sim p \to q) \to p$.

Solution: Since there are two distinct simple statements (two letters) the truth table for this logical statement will have 2^2 or 4 rows. The table is completed below.

p	q	~p	~p→q	(~p→q)→p
T	T	F	T	**T**
T	F	F	T	**T**
F	T	T	T	**F**
F	F	T	F	**T**

Objective B: Programming Applications and the Conditional

Conditional statements are a critical component of a decision-making process. We use conditional statement very often in BASIC programming as well as other programming platforms. For example, we may be interested in executing one part of a program when a certain condition is met and another part of the program when the condition is not met. There are various types of decision structures we can use when programming. In this text, we will focus on the If-Then, If-Then-Else, and the Case structure for programming.

Whenever we depress a key on the keyboard and see the character of that key on the screen, the computer has executed a conditional statement. For example, the computer executes a code of the form: If the key in the J position is depressed, then the output must be the character J. The logical expression executed is of the form IF-Then. In mathematics, we can consider conditional statements such as: if $n=5$, then $n^2 = 25$ or if $x=3$, then $2x-1=5$.

Conditional statements can also be viewed in the form (hypothesis) $\rightarrow$ (conclusion) where one begins with a hypothesis believed to be true and attempt to arrive at a conclusion. This type of representation is more common in a theoretical discipline or in experimentation; however, in this text we will use the conditional as part of a decision process of a program.

Decision Structures in programming:

We outline below the various decision structures we will be using in this text. The specific decision structure that you choose to use in writing a program depends on the problem you are solving and which format you are comfortable with.

The appendix in the back of the text gives a more detailed description on syntax that you may find helpful for each type of decision structure we will mention.

IF-THEN:

Executes only one block of code if a condition is met. If the condition is not met the block of code is not executed and the program terminates.

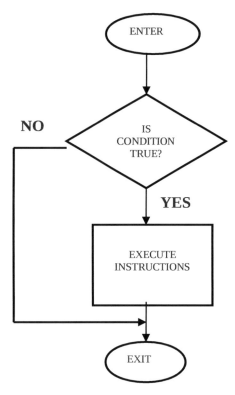

Example 1:   The following program uses two IF – THEN statements to calculate the overtime pay for a person who works over 40 hours per week. You are encouraged to type the program and verify some of the results.

```
REM     This program will calculate weekly salary with
        overtime
REM     Input the number of hours worked and the hourly
        rate
INPUT   "Enter the number of hours worked"; hours
INPUT   "Enter the hourly rate"; rate
IF hours > 40 THEN salary = 40*rate + 1.5*rate*(hours – 40)
IF hours <= 40 THEN salary = hours*rate
PRINT   "hours worked: ";hours, "hourly rate: ";rate,
        "weekly salary: ";salary
END
```

The first IF –THEN statement compares the hours with 40, if it is higher, then the expression **salary = 40*rate + 1.5*rate*(hours – 40)** will be evaluated.
If hours is less than or equal to 40, then the expression **salary = hours*rate** will be evaluated.

If we run the program using 45 hours and 7.50 for the rate, the result will be $356.25

IF-THEN-ELSE: or IF-THEN-ELSEIF:

Executes one of two blocks of code depending on which condition is met.

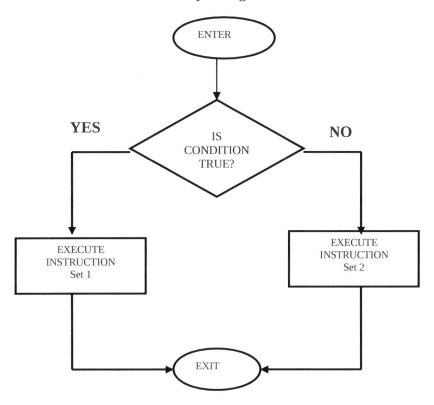

Example 2: The use of IF--- THEN---ELSE is illustrated below:

```
REM     This program will calculate weekly salary with
        overtime
REM     Input the numbers of hours worked and the hourly
        rate
```

```
INPUT     "Enter the number of hours worked"; hours
INPUT     "Enter the hourly rate"; rate
IF hours > 40 THEN salary = 40*rate + 1.5*rate*(hours - 40)
ELSE salary = hours*rate
PRINT     "hours worked: ";hours, "hourly rate: " ;rate,
          "weekly salary: ";salary
END
```

This is just a different way of writing the same program as example 1.

Example 3: The following program makes use of the IF – THEN and ELSE-IF statements.

```
REM       This program will calculate weekly salary with
          overtime
REM       Input the number of hours worked and the hourly
          rate
INPUT     "Enter the number of hours worked" ; hours
INPUT     "Enter the hourly rate" ; rate

IF hours > 40 THEN

              salary = 40*rate + 1.5*rate*(hours - 40)

      ELSEIF hours <= 40 THEN

              salary = hours*rate
END IF
PRINT     "hours worked: ";hours, "hourly rate: ";rate,
          "weekly salary: ";salary
END
```

Case Structure:

There is an alternative to using If-Then and Else-If structure when programming. Although this is not a logical operator, it is a built in code that executes a block of programming code when a condition is satisfied. The benefit in using SELECT CASE structure is when there are more than two program blocks to choose from. This provides a more efficient program.

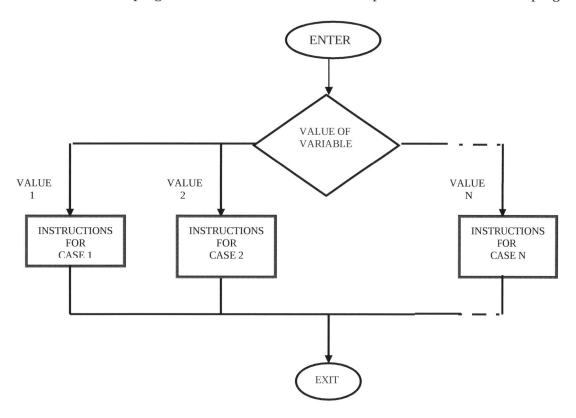

We illustrate the use of SELECT CASE with the salary example including an additional option. We have included the case of someone working more than fifty hours per week. The salary is calculated at straight pay for the first forty hours, time and a half for overtime hours between the next nine overtime hours, and double time for the overtime hours thereafter.

An important point to be made is that BASIC executed line by line. It look for the first condition that is met and executes that line of code and ignores the others. Care must be taken when writing the program so it executes the code you intend. As you can see below, a person working 70 hours will satisfy each case; however, the first line of code is the one that needs to be executed for the program to work properly.

Example 4:   The use of SELECT CASE is illustrated below:

```
REM     This program will calculate weekly salary with
        overtime
REM     Input the numbers of hours worked and the hourly
        rate
INPUT   "Enter the number of hours worked and the hourly
        rate";hours,rate
SELECT  CASE hours
        CASE   IS >=50
                  salary = 40*rate + 2*rate*(hours - 40)
        CASE   IS  > 40
                  salary = 40*rate + 1.5*rate*(hours - 40)
        CASE   IS  < = 40
                  salary = hours * rate
END SELECT
PRINT   "hours worked: ";hours, "hourly rate: ";rate,
        "weekly salary: ";salary
END
```

Note that if we run the program for 52 hours and 10.00 for the rate, the output will be $640.

Objective C: The Biconditional of Statements (Optional)

The biconditional is the last connective we are going to discuss. As the definition and the truth table below indicate, the importance of this logical operator lies in the fact that it allows you to determine when two statements have the same truth values. It does not matter what the truth values of those statements are, but what matters is whether they are the same. There is a nice connection between the conditional and the biconditional that we will discuss in the next section.

Definition: The **biconditional** of two statements p, q, denoted by $p \leftrightarrow q$, read as " p if and only if q" is the statement that is true only when both p, q have the exact same truth values and false otherwise.

p	q	$p \leftrightarrow q$
T	T	T
T	F	F
F	T	F
F	F	T

Example 1: Construct and complete the truth table for the logical statement
$(p \vee q) \leftrightarrow p$

Solution: Since there are two distinct simple statements (two letters) the truth table for this logical statement will have 2^2 or 4 rows. The table is completed below.

p	q	$p \vee q$	$(p \vee q) \leftrightarrow p$
T	T	T	**T**
T	F	T	**T**
F	T	T	**F**
F	F	F	**T**

Example 2: Construct and complete the truth table for the logical statement
$(p \leftrightarrow q) \leftrightarrow \sim p$

Solution: Since there are two distinct simple statements (two letters) the truth table for this logical statement will have 2^2 or 4 rows. The table is completed below.

p	q	$\sim p$	$p \leftrightarrow q$	$(p \leftrightarrow q) \leftrightarrow \sim p$
T	T	F	T	**F**
T	F	F	F	**T**
F	T	T	F	**F**
F	F	T	T	**T**

Objective D: Overview of Truth Tables

This is a a summary of the truth tables covered so far. You may find this page to be a helpful reference for all the logical operators we have discussed this chapter.

Disjunction:

p	q	p ∨ q
T	T	T
T	F	T
F	T	T
F	F	F

Conjunction:

p	q	p ∧ q
T	T	T
T	F	F
F	T	F
F	F	F

Negation:

p	~p
T	F
F	T

Conditional

p	q	p → q
T	T	T
T	F	F
F	T	T
F	F	T

Biconditional:

p	q	p ↔ q
T	T	T
T	F	F
F	T	F
F	F	T

General format for a truth table with three statements

p	q	r	Insert logical expression here
T	T	T	
T	T	F	
T	F	T	
T	F	F	
F	T	T	
F	T	F	
F	F	T	
F	F	F	

EXERCISES 3.5

Construct a truth table for each of the following logical statements:

1. $\sim p \to \sim q$
2. $(p \to q) \land \sim q$
3. $(\sim q \lor p) \to p$
4. $(p \to q) \to r$
5. $(p \leftrightarrow \sim q) \to (p \lor q)$
6. $p \to (q \to r)$

Let p be true, q be false, and r be false, determine the truth value of each of the logical expressions:

7. $(p \land q) \to \sim q$
8. $(p \to q) \to r$
9. $p \to (q \to r)$
10. $(q \to p) \to r$
11. $\sim p \to (p \lor q)$
12. $(p \land \sim q) \leftrightarrow q$

Let p: 4 < 7, q: 8 > 10, and r: 5 > 0. Write each of the following statements as sentences and determine the truth value:

13. $p \to q$
14. $r \to (p \land q)$
15. $(q \lor r) \to \sim p$

What is the output of the program?

17.
```
CLS
INPUT "ENTER A NUMBER"; NUM
IF NUM > 9 THEN
   PRINT "more than 9"
ELSE
   IF NUM <= 9 THEN
      PRINT "Less than or equal to 9"
   END IF
END IF
```

    a) if the number entered is 11?
    b) if the number entered is 0?

18.
```
          CLS
          INPUT "Enter a number"; n
          IF (n > 1) AND (n = 4 OR n < 10) THEN
             PRINT "Hi"
          END IF
          END
```

   a) if the number entered is 1?
   b) if the number entered is 4?
   c) if the number entered is 7?

19.
```
       CLS
       LET message$ = "You can vote in "
       INPUT "How old are you"; age
       IF age>= 18 THEN
          PRINT message$; "the upcoming election."
       ELSE
          PRINT message$; " "; 18 - age; " years."
       END IF
       END
```

   a) if your age is 21?
   b) if your age is 16?

20.
```
          CLS
          INPUT "Enter a grade"; grade
          SELECT CASE grade
             CASE IS > 89
                   PRINT "Your grade is an A."
             CASE IS > 79
                   PRINT "Your grade is a B."
             CASE IS > 69
                   PRINT "Your grade is a C."
        CASE IS >60
          PRINT  "Your grade is a D."
             CASE ELSE
                   PRINT "Your grade is an F."
          END SELECT
          END
```

What is the letter grade if the grade entered is
   a) a 65?   b) a 78?   c) an 89?

COMPUTER ACTIVITY

Student Name_____ Section_____ Date_____

1. Write a program that states the result of the following algorithm:

    a. Choose an integer between 1 and 10.
    b. If the integer is less than 5, then double your integer and add 15, otherwise subtract 4 and square the difference.
    c. State the final result.

2. Jack has a website that sells video games for various platforms. The following price guide is posted where each customer can find what their best purchasing approach is. An order of less than 10 games will cost $20 per unit plus a $7 delivery charge per game. For a purchase of 10 or more games, Jack can charge $15 per unit plus a $5 delivery charge per game.
    a. How much will it cost to purchase 9 games from Jack?

    b. How much will it cost to purchase 15 games?

    c. What are the maximum games you can purchase if your budget is $375?

3. Write a program in BASIC that calculates the results in part a above.

4. Write a program in BASIC to calculate the subtotal (price plus delivery) for Jack's business above.

5. Modify the program so that if 30 or more games are ordered, the delivery charge is given a 50 % discount for each of the excess games.

Section 3.6: Logical Equivalences

It is often necessary to rewrite a logical statement without altering the truth values of the result. We will use this idea later in the section when we construct an electric circuit. We will be able to simplify an electric circuit that contains switches with an equivalent circuit that satisfies the same conditions, but with fewer switches. As discussed in the previous section, the switches will be treated just as statements. This will determine when two logical statements are logically equivalent, and relate that concept to that of a logically equivalent circuit.

Objective A: Logically Equivalent Statements

Definition: Two logical statements p, q are said to be **logically equivalent**, denoted by $p \approx q$ if they each have the exact same truth values in every possible case.

Some textbooks use the notation $p \leftrightarrow q$ to denote equivalence.

Consider the logical statements $p \rightarrow q$ and its truth values in each of the cases. Is there another logical expression with p and q that will result in the same truth values in each of the cases. To help us understand this better, we construct and analyze the truth table for $p \rightarrow q$ and for $\sim p \vee q$.

p	q	$p \rightarrow q$	~p	$\sim p \vee q$
T	T	**T**	F	**T**
T	F	**F**	F	**F**
F	T	**T**	T	**T**
F	F	**T**	T	**T**

Note: The truth values of these statements are identical and therefore we can represent the equivalence by $(p \rightarrow q) \approx (\sim p \vee q)$.

We introduced the biconditional in Section 3.5, but now we can re-state the biconditional of two statements, p, q, in a logically equivalent form: $p \leftrightarrow q \approx (p \rightarrow q) \wedge (q \rightarrow p)$. The truth table below outlines this equivalence.

p	q	$(p \leftrightarrow q)$	$(p \rightarrow q)$	$q \rightarrow p$	$(p \rightarrow q) \wedge (q \rightarrow p)$
T	T	**T**	T	T	**T**
T	F	**F**	F	T	**F**
F	T	**F**	T	F	**F**
F	F	**T**	T	T	**T**

Example 1: Represent $(p \leftrightarrow q)$ with the conjunction and/or disjunction operators only.

Solution: We will make use of two logical equivalences discussed above.
$$p \leftrightarrow q \approx (p \rightarrow q) \wedge (q \rightarrow p)$$
$$(\sim p \vee q) \wedge (\sim q \vee p)$$

We will be able to simplify this logical equivalence further when we cover distributive properties later in this section.

Example 2: Determine whether the two statements below are logically equivalent.
$$p \wedge (\sim q \vee r) \text{ and } p \vee (q \wedge \sim r)$$

Solution:

p	q	r	~q	~r	~q∨r	p∧(~q∨r)	q∧~r	p∨(q∧~r)
T	T	T	F	F	T	**T**	F	**T**
T	T	F	F	T	F	**F**	T	**T**
T	F	T	T	F	T	**T**	F	**T**
T	F	F	T	T	T	**T**	F	**T**
F	T	T	F	F	T	**F**	F	**F**
F	T	F	F	T	F	**F**	T	**T**
F	F	T	T	F	T	**F**	F	**F**
F	F	F	T	T	T	**F**	F	**F**

Since the truth values for the two statements are not the same in each case, the two statements are not logically equivalent.

We now use logical equivalence to verify De Morgan's Laws in logic.

Objective B: De Morgan's Laws

For any two statements p, q, the following statements are true.

1) $\sim(p \vee q) \approx \sim p \wedge \sim q$.

2) $\sim(p \wedge q) \approx \sim p \vee \sim q$

We can verify De Morgan's Laws by examining the truth table of each of the statements. The second law is verified below and the first is left as an exercise.

P	q	(p∧q)	~(p∧q)	~p	~q	~p∨~q
T	T	T	F	F	F	F
T	F	F	T	F	T	T
F	T	F	T	T	F	T
F	F	F	T	T	T	T

Since the columns the statements ~(p∧q) and ~p∨~q have the exact same truth values, the two statements are logically equivalent.

Objective C: Contrapositive, Converse, and Inverse of a Conditional Statement

It may be necessary to occasionally rephrase a statement with another logically equivalent statement. Consider the statement: If I get an A in this course then I will get a car. An equivalent statement is: If I do not get a car, then I did not get an A in the course. This conditional statement can be represented symbolically as follows:

p: I will get an A in the course
q: I will get a car

$p \rightarrow q$: If I get an A in this course, then I will get a car.
$\sim q \rightarrow \sim p$: If I do not get a car, then I did not get an A in this course.

The latter statement $\sim q \rightarrow \sim p$ is called the contrapositive of the original conditional statement. These statements are equivalent, as their truth tables will indicate below.

We can construct other variations of conditional statements called the converse and inverse of the "initial" conditional. You should note however, that the only the "initial" conditional and its contrapositive are logically equivalent. The converse and inverse of an "initial" conditional statement are defined below.

We will begin with a conditional statement $p \rightarrow q$. We will call this our "original" or "initial conditional." The following can be taken as definitions:

1: The contrapositive of $p \rightarrow q$ is $\sim q \rightarrow \sim p$.
2: The converse of $p \rightarrow q$ is $q \rightarrow p$.
3: The inverse of $p \rightarrow q$ is $\sim p \rightarrow \sim q$.

The truth table below outlines the following results:
- The conditional is logically equivalent to its contrapositive.

- The converse and inverse are logically equivalent to each other.

p	q	conditional $p \to q$	contrapositive $\sim q \to \sim p$	converse $q \to p$	inverse $\sim p \to \sim q$
T	T	T	T	T	T
T	F	F	F	T	T
F	T	T	T	F	F
F	F	T	T	T	T

We can summarize the equivalences as follows:
$$(p \to q) \approx (\sim q \to \sim p)$$
$$(\sim p \to \sim q) \approx (q \to p)$$

One should note the converse is in fact the contrapositive of the inverse and vice-versa.

Example 1: State the converse, inverse, and contrapositive of the statement:
If you study hard, then you pass the course.

Solution: Converse: If you pass the course, then you studied hard.
Inverse: If you did not study hard, then you did not pass the course.
Contrapositive: If you did not pass the course, then you did not study hard.

Example 2: State the converse, inverse, and contrapositive of the statement:
If n is an even number, then n^2 is an even number.

Solution: Converse: If n^2 is an even number, then n is an even number.
Inverse: If n is not an even number, then n^2 is not an even number.
Contrapositive: If n^2 is not an even number, then n is not an even number.

Example 3: State the converse, inverse, and contrapositive of the statement:
If $1 < 6$, then $3 >= 5$.

Solution: Converse: If $3 >= 5$, then $1 < 6$.
Inverse: If $1 >= 6$, then $3 < 5$.
Contrapositive: If $3 < 5$, then $1 >= 6$

Note that in the above example, the conditional and the contrapositive are both false while the converse and inverse are both true.

Some other types of useful equivalences include the distributive properties of the conjunction and disjunction. The verification of these properties are left as exercises where a truth table can be constructed and the columns of both statements examined. We know that a necessary condition for the equivalence of those statements can be established if their columns in the truth table are identical in every case.

Distributive properties:

1. $p \land (q \lor r) \approx (p \land q) \lor (p \land r)$
2. $p \lor (q \land r) \approx (p \lor q) \land (p \lor r)$

The reason for part 1 is that if p is true, then ~p is false, and we know that the conjunction of two statements is true only when both statements are true.

The verification of part 2 is left as an exercise.

Objective D: Distributive Properties (Optional)

There are other logical equivalences that we have not yet discussed. This objective will state the distributive properties as they apply to logical statements. Back in Chapter 1 we discussed distributive properties of real numbers. The distributive properties here are similar, however we distribute statements rather than real numbers. When designing digital circuitry, one may wish to achieve a certain output based on a certain input with the least amount of components. Distribution and simplification properties can help us in our design.

Distributive properties include

$$p \land (q \lor r) \approx (p \land q) \lor (p \land r)$$
$$p \lor (q \land r) \approx (p \lor q) \land (p \lor r)$$

The first distributive property is often referred to as the distributive property of the conjunction over the disjunction. The second property is referred to the distributive property of the disjunction over the conjunction.

Other simplification properties include:

$$p \lor \sim p \approx T$$
$$p \lor F \approx p$$
$$p \land F \approx F$$
$$p \land \sim p \approx F$$

The above simplification properties assume that if p is a statement, the result will be true or false; however, as discussed earlier, if p denotes the position of a switch, the result of the logical operator will be a 0 or a 1. More will be discussed about electric circuits in the next section.

Example: Use a distributive property to simplify the statement $p \wedge (q \vee \sim p)$
  Solution:
$$p \wedge (q \vee \sim p)$$
$$\approx (p \wedge q) \vee (p \wedge \sim p)$$
$$\approx (p \wedge q) \vee F$$
$$\approx (p \wedge q)$$

We will not go into any more detail about distributive and simplification properties here. The intention of this objective is to make the reader aware that there are further properties in logic that are not discussed in detail in this text. A pure logic textbook discusses these and other concepts in more detail. In the next section, Section 3.7, we will further discuss electric circuits that were mentioned here.

EXERCISES 3.6

Use the truth table to determine if the following statements are logically equivalent.
1. $\sim (p \vee q)$ and $\sim p \wedge \sim q$
2. $p \wedge (q \vee r)$ and $(p \wedge q) \vee (p \wedge r)$
3. $p \to q$ and $\sim p \wedge q$
4. $p \to (q \vee r)$ and $(p \to q) \vee (p \to r)$
5. $\sim (p \to q)$ and $p \wedge \sim q$

Use the De Morgan's Law to rewrite the following negations:
6. It is false that roses are red or the leaves are green.
7. NOT(3 > 5 and 8 < 12)
8. NOT ($x = 4$ OR $y < 10$)

For each of the following conditional statements, write the contrapositive, converse, and inverse:
9. If you get a ticket, then you violated the traffic rules.
10. If today is Saturday, then it is raining.
11. If it has wings, then it flies.

12. If your GPA is 3.5 or higher and you carry a twelve-credit load, then you are on the Dean's List.

For each of the following conditional statements, write the contrapositive, converse, and inverse. Also, find the truth value for each statement:

13. If 2 < 4, then 3 > 5.
14. If 6 > 8, then 9 < 11.
15. Show that $(p \vee \sim p)$ is always a true statement.

Section 3.7: Applications with Electric Circuits

The construction of electric circuits is common in computer architecture. The most simple type of circuit you may think of is one with a battery source (the input) and a light bulb (the output) and a switch. If the switch is closed, then the light will be on and if not, it will be off. The output of the circuit depends on whether the switch is open or closed.

Below we illustrate a simple circuit.

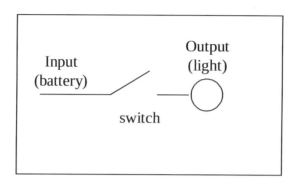

Today, circuitry is everywhere around us. It is in the form of logical circuit boards and they are found in everything from washing machines, microwaves, telephones, televisions, and of course computers. Although actual switches are uncommon, the use of digital circuitry performing the same operation is used instead. The switches are referred to as gates consisting of millions of transistors and can be found in chips less than one inch in length. Other common types of electronic components that are used in circuitry include resistors, capacitors, inductors, and power supplies. In fact, there are many pre-designed circuits in the form of a computer chip that can also be used as a component in a circuit. You can think of this as a circuit connected with another circuit. For simplicity in illustrations, we will use switches with the understanding that a simple extension of these concepts can lead to many more types of possible electric circuits.

A nice way to think of these circuits is to imagine current flowing in from the input, and depending on the set up of the switches, current may or may not flow to the output. You may also think of this as a bridge connecting two pieces of land. If the bridge is open, you cannot get from where you begin (the input) to where you may want to go (the output).

The types of circuits we will be discussing fall into one of the following categories.

Category 1: A <u>Series Circuit</u>: There is only one path for current to flow. You may think of this as only one path to walk.

Category 2: A <u>Parallel Circuit</u>: There is more than one path for current to flow. You may think of this as walking from one place to the other and you may get there by following more than one path.

Category 3: A <u>combination of series and parallel circuits</u>: At times, there is only one path to follow, while at others you may have a choice of more than one path to follow.

There will be two general rules we will use to construct electric circuits from logical statements.

1. A series circuit with two switches p, q, will be described by the conjunction of p and q, $p \wedge q$. The reason for this is that it will be necessary for both of the switches to be closed for current to flow from the input to the output. The electric circuit corresponding to $p \wedge q$ is therefore,

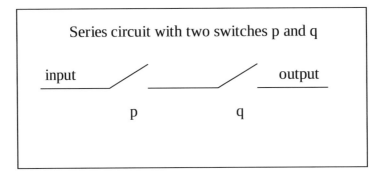

Consider the following Truth Table and notice that we have changed the possible values from T to 1 and from F to 0. We can further identify the "1 to indicate a closed switch" and the "0 to indicate an open switch".

p	q	$p \wedge q$
1	1	1
1	0	0
0	1	0
0	0	0

The truth table illustrates all possibilities for the position of the two switches. Either both are closed, p is closed and q is open, p is open and q is closed, or both are open.

You should note that current would flow from the input to the output only when p and q are both closed. That is, when the conjunction of p, q is true; in this case a 1.

2. A parallel circuit with two switches p, q, is described by the logical statement $p \vee q$. This is because it is necessary for one or both of the switches to be closed for current to flow from the input to the output. The truth table that illustrates all possible positions for these switches and the corresponding output is therefore,

p	q	$p \vee q$
1	1	1
1	0	1
0	1	1
0	0	0

And the circuit is of the form:

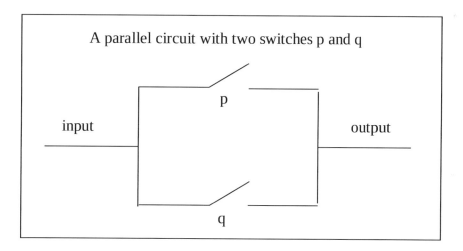

A parallel circuit with two switches p and q

Note: It is common to show the switches in the open position. This is simply the notation of a switch and the analysis of the circuit should follow from all possibilities from the truth table.

Example: Construct the electric circuit for $p \vee (q \vee \sim r)$.

The electric circuit is a parallel circuit with two branches. The first branch contains switch p, while the other branch contains switches q, ~r connected in parallel. The diagram for this electric circuit is then,

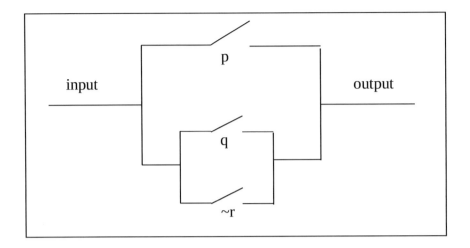

EXERCISES 3.7

Construct the truth table and electric circuit diagram for the given logical statements. Then describe the behavior of the circuit with all various possibilities and list the state (open or closed) for each of the switches when there is current flow from the input to the output.

1. $(p \wedge q) \vee r$
2. $\sim p \wedge (r \vee q)$
3. $[(p \wedge q) \vee \sim r] \wedge (p \vee r)$
4. $p \rightarrow q$ (Hint: the equivalence $p \rightarrow q \approx \sim p \vee q$ may be useful)
5. $(p \wedge q) \vee (r \wedge \sim p)$
6. $(r \vee q) \wedge (r \vee \sim p)$
7. $\sim (p \vee \sim q) \vee (q \wedge r)$ (Hint: De Morgan's Laws may be useful)
8. $[(p \vee \sim q) \vee (q \wedge r)] \vee (p \rightarrow q)$
9. $(\sim p \rightarrow q) \rightarrow (r \wedge \sim q)$
10. $p \leftrightarrow q$ (Hint: the equivalence $p \leftrightarrow q \approx (p \rightarrow q) \wedge (q \rightarrow p)$ may be useful)

Chapter 4

Relations, Functions, and Subroutines

In this chapter, we introduce several new and related concepts that will enable us to organize programs more effectively. Relations, functions, and subroutines are vital concepts in computer technology. They provide us with a foundation for creating programs that are more efficient and easier to read and follow.

We begin by defining and explaining the relation concept. All computer programs are written to solve problems defined by relations. Along with relations we discuss the related idea of functions. The function concept is critical in mathematics. We use the mathematical description of a variety of functions to gain a deeper understanding of their use and also show how they relate to BASIC.

We then introduce the computer concepts of subroutines and repetition structures. These will expand our capability to write more complicated and useful programs.

After completing this chapter you should have obtained a good understanding of relations, functions, and subroutines. You should come away with a greater appreciation for how they can be used to further enhance the computational capabilities of a computer.

More specifically, you should be able to:
- understand what relations and functions are and be able to distinguish between the two,
- create a solution strategy to solve problems,
- work with function notation,
- perform basic operations with absolute value, MOD, exponential and logarithmic functions,
- understand and build subroutines and functions in BASIC,
- understand and work with the three repetition structures in BASIC,
- understand what recursion and recursive functions are.

Section 4.1: Relations and Functions

An IT professional may be asked to design a computer network within a fixed budget. Obviously, there is a relationship between the budget and the type of computer system that can be designed. In fact, for any given budget there is an almost limitless set of configurations that can be produced.

The health level of a player in an action/adventure video game depends upon how often the player is wounded in battle. There is a relationship between the health level and the number of hits a player takes.

A problem with an Internet connection may have several possible causes. Consequently, the problem can give rise to several possible solutions. There is a relationship between the problem and its possible solutions.

What do all these examples have in common? They all involve a relationship between items of interest. Relations are not just an abstract mathematical concept. We work with relations every day. In this section we introduce the notion of relations along with the closely associated concept of functions.

Objective A: Defining Relations

A relation involves an association between two groups of items or conditions. Given one group, called the input, the relation produces the other group, the output. For a given input there are one or more outputs. To understand and work with relations better, we look at them more closely. We start by defining what a relation is.

Definition: A **relation** is simply a rule (correspondence) that takes an input and produces one or more outputs.

This can also be illustrated graphically as in the figure below. The input is transformed by the relation into one or more outputs.

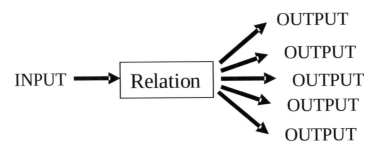

To further illustrate what a relation is, we introduce the following examples:

Examples:

1.

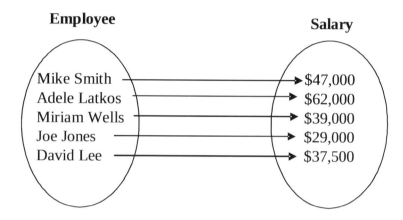

The input is the employee and the output associated with each employee is his or her salary. The relation is the correspondence between the employee and their salary.

2.

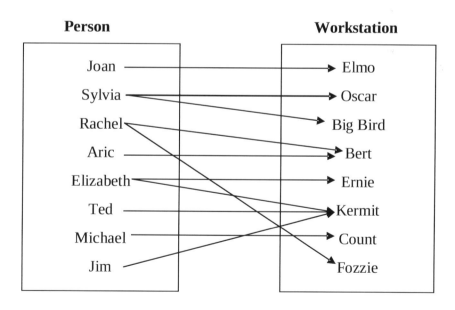

The input is a person's name and the output associated with each person is the computer workstation they use. The relation is the correspondence between the person and the workstation or workstations they use.

3.

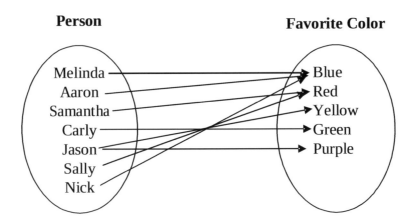

The input is a person's name, and the output associated with each individual is their favorite color or colors. The relation is the correspondence between the person and a color or set of colors.

Some relations can also be expressed mathematically.

Examples:

4. Consider the positive square root of a number. The square root of 4 is 2, i.e.,

$$\sqrt{4} = 2, \text{ since } 2^2 = 4.$$

This is an example of a relation. The input 4, is related to one output, the positive number 2.

5. Suppose now that we wish to find any number whose square is equal to 4. There are two answers here; namely, 2 and –2. That is to say, 2 and –2 are two different square roots of 4. Again, this is a relation with two different outputs, 2 and –2, for one input of 4, i.e.,

$$\pm\sqrt{4} = \pm 2, \text{ since } (-2)^2 = 4 \text{ and } (+2)^2 = 4$$

We can also represent this relation using a table:

INPUT	OUTPUT 1	OUTPUT 2
4	+2	-2
9	+3	-3
16	+4	-4
⋮	⋮	⋮

The inputs are numbers we want to find the square root of, and the outputs are the positive and negative square roots of these numbers. Again we note that each input in this relation, has two outputs.

Our ability to see and understand relations is directly related to our ability to understand and solve problems. A good IT professional is capable of coming up with many possible solutions to a problem. That is, he or she can devise a relation that links the input (a problem), to the output (a set of possible solutions). Thus, the better you are at seeing and creating relations, the better prepared you'll be at solving problems. People who possess this skill are referred to as "nonlinear" thinkers and are in great demand.

Objective B: Defining Functions

Functions are a special type of relation that help in making programs more organized and easier to write and understand. In this sub-section we take a closer look at functions and their implementation in BASIC.

We know that any process that takes an input and produces an output is a relation. A function is a particular type of relation. Any relation that yields only one output for any given input is called a **function**. Functions are one of the most important concepts in mathematics and we will show that they have a large impact on computers as well.

We begin with the definition:

Definition: A **function** is a rule (correspondence) that takes an input and produces only one output.

Consider an IT professional who has designed a network with 4 workstations, a printer and a router for a business office. The cost of the workstations, printer and router are all fixed and total $12,000. The cost of the cable varies. Thus, the cost of our new network is simply the fixed costs of the electronic devices, plus the total length of the cable. If x denotes the total length of the cable needed, and the cable can be purchased for $75 per foot, then the total cost in dollars is given by:

Total Cost in Dollars = 12,000 + 75x.

Here we would say the Total Cost is a function of the length of the cable. Given an input, the cable length, you get one and only one output, the total cost of the network.

This function can also be represented in table form:

INPUT Cable Length	OUTPUT Total Cost
0	$12,000
20	$13,500
100	$19,500
⋮	⋮

Note: There is only one output for each input of cable length.

Consider another example. The formula for changing Celsius temperature to Fahrenheit temperature is given as

$$F = \frac{9}{5}C + 32.$$

The Fahrenheit temperature, F, is a function of the Celsius temperature, C. For any given Celsius temperature, C, we have one associated Fahrenheit temperature, F. This is an equation involving two variables (F and C). We know that at sea level, water boils at 100°C. At what temperature Fahrenheit is this? Substituting in 100 for C we obtain the equivalent Fahrenheit temperature as:

$$F = \frac{9}{5}(100) + 32 = 9(20) + 32 = 180 + 32 = 212°.$$

Note: All functions are relations, but not all relations are functions.

Examples: Determine which relations are functions.
1. cost of a stamp

    1990 ⟶ 20
    2001 ⟶ 34
    2002 ⟶ 37

2. positive square root of a number

    $\sqrt{4}$ ⟶ 2
    $\sqrt{9}$ ⟶ 3
    $\sqrt{36}$ ⟶ 6

3. absolute value

    $|-4|$ ⟶ 4
    $|4|$
    $|-6|$ ⟶ 6

Solution:
1. Is not a function, since the year 2002 goes to two different outputs: 34 and 37.
2. Is a function, since each input goes to one and only one output.
3. Is a function, since each input goes to only one output. Note that two different inputs can go to the same output, though.

Objective C: Functional Notation

Functions have a special notation that we will use both mathematically and in programming. This notation provides a convenient way to reference and work with functions. We illustrate this with an example.

Suppose that we wish to define the positive square root function. Mathematically we would represent it with the square root symbol. To signify that it is a function, we write it in a way that allows us to identify the input and also assign it a "name." This enables us to work with functions in a more general way, e.g.,

$$f(x) = \sqrt{x}, \ x \geq 0$$

Here:

1. *f* is the **name** of the function.

2. *x* is called the **argument** of the function; that is, the operation *f* is performed on *x*. Functions may have one argument, several arguments... or no arguments at all. We will refer to the argument as the **input** of function *f*.

3. *f(x)* read "*f* of *x*" denotes the result of the function *f* when it is applied to *x*. We will refer to this result as the **output** of function *f*.

4. The right-hand side of this definition (that is, the expression immediately to the right of the equal sign) denotes the operations to be performed on the arguments. Here, we wish to take the square root. Since only nonnegative values result from this function, we say that the output or **range** of *f* is the set of all nonnegative real numbers.

5. The condition $x \geq 0$ is used, since the square root function is defined only over the set of nonnegative real numbers. The "x" values are the inputs, thus we say that the input or **domain** of *f* is the set of all nonnegative real numbers.

NOTE: In programming, some functions allow for different numbers of arguments or have optional arguments.

This notation provides a convenient way to evaluate and work with functions as illustrated in the following examples:

Examples: Evaluate each function at the indicated values:
1. $f(x) = 2x - 3$, find $f(-1)$
2. $g(t) = -t^2 - 3t + 1$, find $g(3)$
3. $h(c) = \sqrt{3c + 3}$, find $h(2)$

Solutions:
1. This notation means, given the function defined by *f(x)*, evaluate it when $x = -1$ i.e. $f(-1) = 2(-1) - 3 = -5$ [We just substituted −1 into the formula in place of x.]

2. This notation means, given the function defined by g(t), evaluate it when t =3, i.e. $g(3) = -(3)^2 - 3(3) + 1 = -9 - 9 + 1 = -17$
3. Similarly, $h(2) = \sqrt{3(2)+3} = \sqrt{6+3} = \sqrt{9} = 3$

Examples: Given $f(x) = \sqrt{x}$, $g(x) = x^2 - 1$ and $h(x) = |x| - 2$, evaluate the following:
1. f(9)
2. g(-1)
3. h(-5)
4. f(-1)
5. f(2)
6. g(5)
7. h(-2)

Solutions:
1. This notation means, given the function defined by f(x), evaluate it when x = 9, i.e. $f(9) = \sqrt{9} = 3$
2. This notation means, given the function defined by g(x), evaluate it when x = −1, i.e. $g(-1) = (-1)^2 - 1 = 1 - 1 = 0$
3. $h(-5) = |-5| - 2 = 5 - 2 = 3$
4. $f(-1) = \sqrt{-1}$ is undefined, since −1 is not in the domain of the square root function, since we cannot take the square root of a negative number.
5. $f(2) = \sqrt{2} \approx 1.4142K$
6. $g(5) = (5)^2 - 1 = 25 - 1 = 24$
7. $h(-2) = |-2| - 2 = 2 - 2 = 0$

Example 8 The formula for changing Celsius temperature to Fahrenheit temperature was given above as:
$$F = \frac{9}{5}C + 32.$$
Write this as a function using function notation and then find the Fahrenheit temperatures associated with the following Celsius temperatures: 0°C, 100°C, -5°C, 32°C, and -20°C.

Solution: We can rewrite the formula to show the functional form as:
$$F(C) = \frac{9}{5}C + 32$$

To determine the Fahrenheit temperature F for the given Celsius temperatures, we evaluate each of the following:

$F(0)$, $F(100)$, $F(-5)$, $F(32)$, and $F(-20)$.

$$F(0) = \frac{9}{5}(0) + 32 = 32$$

$$F(100) = \frac{9}{5}(100) + 32 = 9(20) + 32 = 180 + 32 = 212$$

$$F(-5) = \frac{9}{5}(-5) + 32 = 9(-1) + 32 = -9 + 32 = 23$$

$$F(32) = \frac{9}{5}(32) + 32 = \frac{288}{5} + 32 = \frac{288}{5} + \frac{160}{5} = \frac{448}{5} = 89.6$$

and

$$F(-20) = \frac{9}{5}(-20) + 32 = 9(-4) + 32 = -36 + 32 = -4$$

A computer program can also be viewed as a function, since most programs take an input and produce a single well-defined output. The program/function contains the rule or process that shows how an input is transformed into the output.

EXERCISES 4.1

1. What is a relation?

2. What is a function?

3. Are all functions also relations? Explain.

4. Are all relations also functions? Explain.

5. What is the domain and range of a function?

6. Create a simple relation? Is your relation also a function? Explain.

7. You have been asked to design a computer network for an office space that is 20 feet by 40 feet in area. You need to set up 4 computer workstations that can be placed anywhere in the office, however, each work area must have at least 40 square feet of space. You will also need one meeting room that has 200 square feet of space. In addition you must include one conveniently located printer and all the computers will go through a single router that accesses the Internet. It costs $1 per foot of cable and the cable can either go along the base of the walls (there must be at least two entry/exit doors) or above in the ceiling. The ceiling height is 8 feet. Each workstation costs $2,500, the printer costs $1,000 and the router is $1,000. You have been given a budget of $12,500. Design two different office layouts and determine the costs of each system.

8. You are asked to set up a wireless network in an office area as shown below. Each wireless transmitter/receiver has an effective radius of coverage of 25 feet. Design at least two wireless networks that cover 100% of the area. Sketch your results on paper. [Helpful hint: you may find a compass (a device for drawing circles) useful.]

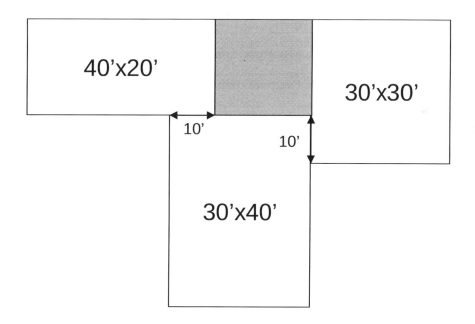

Identify which relations are functions

9. a.

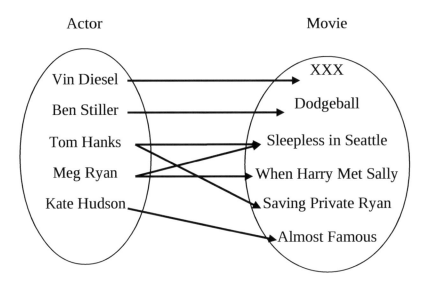

b. A number squared

2 ⟶ 4
4 ⟶ 16
5 ⟶ 25

c. Number of siblings

Tom ⟶ 2
Sue ⟶ 3
Danise

Evaluate each function at the indicated values

10. $f(x) = -2x + 5$, Find
    a. $f(2)$
    b. $f(-3)$

11. $g(x) = -2x^2 + 5x - 3$, Find
    1. $g(0)$
    2. $g(-1)$

12. $f(x) = \dfrac{x^2 - 1}{x - 2}$, Find
    a. $f(0)$
    b. $f(2)$
    c. $f(-2)$

13. $g(x) = \dfrac{1}{3}x^2 - x$, Find
    a. $g(3)$
    b. $g(1)$

14. The formula for changing Fahrenheit temperature to Celsius temperature is $C = \dfrac{5}{9}(F - 32)$. The Celsius temperature, C, is a function of the Fahrenheit temperature, F.

a. Write this function using function notation.
b. Find the Celsius temperature that corresponds to the following Fahrenheit temperatures: 0°F, -10°F, 82°F, 212°F, 32°F, and 29°F.

15. The formula for simple interest earned for a period of 1 year is given as: $I = pr$, where p is the principal invested and r is the simple interest rate. If $10,000 is invested, the simple interest earned in 1 year as a function of the interest rate r is, $I(r) = 10,000r$. Determine the simple interest earned for each interest rate r: $r=4.5\%$, $r=7.25\%$, and $r=19.7\%$.

16. The monthly interest charge on a credit card is given as: $I = p\dfrac{r}{12}$, where p is the principal owed on the card and r is the yearly interest rate. If $6,000 is owed on the card then the monthly interest owed on the card as a function of the interest rate r is: $I(r) = 6,000\dfrac{r}{12} = 500r$. Determine what the minimum payment needs to be so that the amount owed does not increase on a card with $r=3.99\%$, $r=7.99\%$, $r=18.99\%$, and $r=23.99\%$

17. In an action adventure computer video game, the health level of a player is computed using the function: $H(x) = 100 - \dfrac{x}{5}$, where x is the number of times the player is hit by another player. The player's health is rated: Excellent if $H \geq 85$, Good if $65 \leq H < 85$, Fair if $40 \leq H < 65$, Poor if $0 < H < 40$ and Spirit if $H \leq 0$.

a. Determine the health level and rating of the player for each of the following: $H(45)$, $H(450)$, $H(212)$, $H(625)$.
b. Determine the number of shots required before a players health level changes from Excellent to Good to Fair to Poor to Spirit.

Section 4.2: Some Special Mathematical Functions

BASIC has a number of built-in functions, some of which we will explore in more depth throughout this chapter. In this section we present some of the more common types of functions that the computer professional should be familiar with. We show both the mathematical and the BASIC representations of these functions.

Objective A: Square Root Function

We are already familiar with the square root function from Chapter 1, although at that time we did not refer to it as a function. Instead, we called it an operator. In general, functions are also operators. As we mentioned in the last section, the square root function is defined mathematically as:

$$f(x) = \sqrt{x}, \ x \geq 0$$

BASIC Form of the Square Root Function

SQR(x): Returns the principal square root of a numeric expression if it exists. The principal square root is the positive square root of the number.

BASIC and mathematical notation: $\text{SQR}(x) = \sqrt{x}$ if $x \geq 0$
Domain: $x \geq 0$
Range: $\text{SQR}(x) \geq 0$

BASIC returns the error "Illegal Function Call" if SQR is invoked using a negative argument. Why is that? (See example 5 below, for a brief explanation.)

Examples:
1. SQR(16)=4
2. SQR(2)=1.414214…
3. SQR(999)=31.60696…
4. SQR(0)=0
5. SQR(33-48) gives an error because this is equivalent to SQR(-15) and the square root of a negative number is not defined. Negative numbers are not allowable inputs into the square root function. They are not in the domain of the function.

Objective B: Absolute Value Function

The absolute value function is defined mathematically as:

$$f(x) = |x| = \begin{cases} x & x \geq 0 \\ -x & x < 0 \end{cases}$$

BASIC Form of the Absolute Value Function

ABS(x): Returns the absolute value of a numeric expression.

BASIC and mathematical notation: $\text{ABS}(x) = \begin{cases} x \text{ if } x \geq 0 \\ -x \text{ if } x < 0 \end{cases}$

Domain: $x \in R$ [x is any real number]
Range: $\text{ABS}(x) \geq 0$

In a sense, this function "ignores" the sign of the argument and returns just the numerical part of it as illustrated in the examples below.

Examples:
1. ABS(123)=123
2. ABS(-14.56)=14.56
3. ABS(0)=0
4. ABS(33-48)=ABS(-15)=15

Objective C: MOD Function

We've discussed the MOD function in section 1.6. Here we take a closer look at how we use it in BASIC.

BASIC Form of the MOD Function

x MOD y: Returns the remainder of $x \div y$

BASIC and mathematical notation: $x \text{ MOD } y = \text{remainder}\left(\frac{x}{y}\right), \text{ if } y \neq 0$

Domain: $x, y \in R, y \neq 0$
Range: $x \text{ MOD } y \in \{0, 1, \ldots, y-1\}$

This function is not written as MOD(x,y), as we might expect. While the functions we have considered thus far have been created and invoked using **prefix** format, this function takes the form of an **infix** operator, like the +, or addition operator. Like addition, the operands are on both sides of the MOD operator.

Examples:
1. 17 MOD 3=2
2. 18 MOD 3=0

BASIC returns the error "Division By Zero" if the second operand of MOD is zero.

Example 3: 17 MOD 0 gives an error

Non-integer operands are rounded to the nearest integer before the operation is performed.

Example 4:   15.2 MOD 3.7=3

Note: It is possible to take the modulus of negative numbers. However, we will not discuss this, since it is beyond the scope of what we want to cover.

Objective D: Exponential Functions

The exponential function is a very important function in mathematics and computer science. Understanding the exponential function and its properties is critical to a number of related computer concepts.

To understand the exponential function, we start with the following type of problem. We start by picking a number. To make things easier we take a simple number and pick the number 2.

The first step is to multiply it by 2 to obtain:	2 * 2 = 4
Next, we multiply that result by 2 to get:	2 * 2 * 2 = 8
Multiply the result by 2 again:	2 * 2 * 2 * 2 = 16
Multiply the result by 2 again and we have:	2 * 2 * 2 * 2 * 2 = 32.

Notice how fast the number grew at each stage. The larger the number grew, the faster it grew. Continue multiplying the number by 2 a few more times, to get a feel for how rapidly it grows.

This is the fundamental property that makes an exponential function different from any other function. The growth rate of an exponential function is directly proportional to its value.

Looking at our example above, the first multiplication by 2, when the value was 2, increased the result to 4 for a net increase of 2. But later when its value had increased to 16, the same process of multiplying by 2 increased it by 16 to 32. This shows that the rate of growth depends on the current value, since we always multiply the current value by 2.

Now looking back at the equations above, we see that we can also write them using our exponential notation.

First, we have:	$2 = 2^1$
Then we multiply by 2 to get:	$4 = 2^2$
Then multiply by 2 a second time:	$8 = 2^3$
A third:	$16 = 2^4$
A fourth:	$32 = 2^5$

We can see that a general pattern emerges: at the nth multiplication by 2 we'd have 2^n of them.

Of course, we can do the same thing with multiplying by 3's instead of 2's, or we could take half of the number each time instead (i.e., multiply by ½). In fact, we can choose any number, other than 1, and just keep multiplying by that number. We call this number "a." Then after the nth step we get the number a^n, that is "a" multiplied by itself n-times.

Suppose we want to increase our number gradually, instead of using integer steps. This could be more useful in some problems where we know something is increasing steadily. For example, let's say we have a problem with a quantity that doubles every hour. We may want to know how much has it grown after an hour and a half. Then between the first and second steps we'd want to get a result that's bigger than 2, but not as big as 4. So instead of 2^1 (after 1 hour) or 2^2 (after 2 hours) we just use $2^{1.5}$ to find the value after 1.5 hours. Most of us don't know the value of $2^{1.5}$, but if we use a calculator we can find the value:

$$2^{1.5} = 2.8284...$$

From the rules of exponents in Chapter 1, we know that we can have any real number as an exponent, so in this way we can calculate the value at any "x," which is just 2^x.

This is called an **exponential function** of "x," because the "x" is the exponent and is written, in general, as:

Definition: An **exponential function** is any function of the form:

$$f(x) = a^x,$$

where a is any number such that a > 0, a ≠ 1, and x is any real number.

This is read "*f* of x equals a to the x." The number "a" is called the **base** of the exponential function.

Example 1: Write an exponential function with base a=4.

  Solution: Since a = 4 we replace "a" in the definition to obtain: $f(x) = 4^x$

Example 2: Write an exponential function with base a=1/2.

  Solution: Since a = 1/2 we replace "a" in the definition to obtain: $f(x) = \left(\dfrac{1}{2}\right)^x$

The exponential function is different from power functions that you may have seen before. You should recall that power functions are functions where you have a variable, x, and raise it to a constant power. For example, consider the quadratic function: $f(x) = x^2$ where the variable, x, is the quantity that's being raised to the 2nd power. For exponential functions the variable is in the exponent.

The Natural Exponential Function

The base of the exponential function can be any positive real number except 1. A very important number in mathematics is the irrational number "e". The number "e" arises quite naturally in a number of areas of mathematics, such as population growth and continuous compounding of interest.

The exact value is represented using the letter "e", but if we need to compute with it, we use the decimal approximation e ≈ 2.71828...

Definition: The function

$$f(x) = e^x,$$

is called the **natural exponential function**.

The natural exponential function has many different application areas: mathematics, nuclear physics, and finance are just a few of them.

> **ASIDE**: When computing the continuous compounding of interest we start with the compounding interest formula:
>
> $$A = P\left(1 + \frac{r}{n}\right)^{nt}.$$
>
> Here P is the principal amount of the investment, r is the interest rate (given as a decimal), n is the number of compoundings per year, t is the number of years that P is invested, and A is the amount the investment grew to after t years. If we allow the number of times we compound per year, n, to get very large, then it is possible to show that
>
> $$A = Pe^{rt}, \text{ and}$$
>
> $$e \approx \left(1 + \frac{1}{m}\right)^m \text{ where } m = \frac{n}{r}.$$
>
> For example, if we evaluate this expression with m=10,000 we would replace m with 10,000 and compute
>
> $$\left(1 + \frac{1}{10,000}\right)^{10,000} \approx 2.718146,$$
>
> which surprisingly is a good approximation to $e \approx 2.71828...$

Example 3: Given the natural exponential function $f(x) = e^x$, find $f(0), f(1), f(-1)$ and $f(2)$

Solution: We just evaluate the function at $x = 0, 1, -1$ and 2 respectively, i.e.
$f(0) = e^0 = 1$
$f(1) = e^1 \approx 2.71828L$
$f(-1) = e^{-1} \approx 0.367879L$
$f(2) = e^2 \approx 7.38905L$

BASIC Form of the Exponential Function

In BASIC the command EXP(x) returns the value of e^x, where e is the base of the natural exponential function.

BASIC and mathematical notation: $\text{EXP}(x) = e^x$
Domain: $x \in R$
Range: $\text{EXP}(x) \geq 0$

Examples:
  4. EXP(0)=1
  5. EXP(1)=2.718282…

Objective E: Properties of Exponential Functions (Optional)

Let's consider the exponential function from the example in Objective D above:

$$f(x) = 2^x$$

Now let's ask the question, what happens as we vary the exponent x?

First as x gets larger, 2^x gets larger, since

$$2^1 < 2^2 < 2^3 L$$

$$2 < 4 < 8 L \ .$$

What happens when x=0?

We know from our properties of exponents in Chapter 1 that, when x=0, then $2^0=1$.

What happens if x becomes negative?

Consider the following:

when $x = -1$, $f(-1) = 2^{-1} = \dfrac{1}{2}$

when $x = -2$, $f(-2) = 2^{-2} = \dfrac{1}{2^2} = \dfrac{1}{4}$

when $x = -3$, $f(-3) = 2^{-3} = \dfrac{1}{2^3} = \dfrac{1}{8}$

Thus we can see that as x becomes more negative (decreases) then $f(x) = 2^x$ also decreases.

We see from above that as x gets more negative, $f(x) = 2^x$ gets smaller, but no matter how small x gets, $f(x) = 2^x$ is never zero. Thus, the function 2^x is never zero for any value of x.

These four properties hold for any exponential function of the form $f(x)=a^x$, where $a > 1$, i.e.

1. $f(x)=a^x$, increases when x increases.
2. $f(x)=a^x$, decreases when x decreases.
3. $f(x)=a^x$, equals one (1) when $x=0$, i.e. $f(0)=a^0=1$
4. $f(x)=a^x$, never equals zero (0), for any value of x.

To illustrate just how rapidly an exponential functions grows, let's compare it to some other more familiar functions. This is shown in the table below:

x	4x	x^2	3^x
0	0	0	1
5	20	25	243
10	40	100	59,049
100	400= 4.0 x 10^2	10,000=1.0 x 10^4	5.1538 x 10^{47}
1000	4,000= 4.0 x 10^3	1,000,000=1.0 x 10^6	Larger number than most calculators can display. You may get an error or overflow message.

Objective F: Logarithmic Functions

What are logarithmic functions and why should we study them? Have you ever heard of a report of an earthquake that measures 6.8 on the Richter scale? Do you know what that means? Did you know that an earthquake that measures 7.8 on a Richter scale is 10 times more powerful than a 6.8, or that an 8.8 is one hundred times more powerful and that a 9.8 is **one thousand times more powerful**? An understanding of logarithmic functions will show us why.

Also, you have probably used search engines on the Internet. How do they know which method to use and which ones are faster? It turns out that the knowledge of logarithms is also useful here. In this section we will look at logarithmic functions so that we can answer these questions.

Definition of Logarithms:

Logarithmic functions are very closely related to exponential functions. As we have just shown, the exponential function takes a number, "a," called the base and raises it to a power x to obtain another number b, i.e.,

$$b = a^x$$

The logarithmic function on the other hand, works in reverse. It starts with the number "b" and converts it to "x," such that $b=a^x$. We express this in the following way:

Definition: The **logarithm** of x in **base** a, written as

$$\log_a(x)$$

is defined for all positive numbers "a", where $a \neq 1$, and

$$y = \log_a(x) \text{ means } x = a^y.$$

For example, if we are asked to find $\log_2(8)$, then we are really solving the related problem of finding x such that:
$$8 = 2^x$$

This answer is $x = 3$, since $2^3 = 8$.

The process of determining the exponent "x," such that the above expression is true, is called finding the logarithm.

Simply put, logarithms are exponents!

One thing we can see is that there are different logarithms, depending upon which base, "a," we are considering. Obviously we would get a different value for x if the base a=2 as opposed to a=3.

Understanding logarithms can be quite difficult. Let's look at some examples to become more familiar with it.

Example 1: Find the base 2 logarithm of 16, written as $\log_2(16)$.

Solution: We are looking for the value of "x" that makes the following statement true:

$$2^x = 16,$$

i.e., how many times do I multiply the base 2 by itself to get 16.

Well 2*2=4, 2*2*2=8 and 2*2*2*2=16. Thus, the answer is 4 times.

Thus,

$$\log_2(16) = \log_2(2^4) = 4$$

The number 4 is just the exponent of the corresponding exponential term 2^4.

This is why logarithms are called exponents.

Example 2: Find $\log_4(16)$.

Solution: We must solve the related problem:

$$4^x = 16,$$

i.e., how many times do I multiply 4 to get 16. In this case it is easy since 4*4=16, the answer is 2.

Thus,

$$\log_4(16) = \log_4(4^2) = 2$$

Can you solve the following?

$$\log_{10}(100)$$

Again, we follow the same procedure and solve the related exponential equation:

$$10^x = 100$$

The answer is 2 since $10^2 = 100$.

These examples are all straightforward, since the exponents were always positive exponents. We only have to look at the base of the logarithm and then figure out how many times we must multiply the number by to get the desired result. Let's extend the result to slightly more complicated numbers.

Logarithms can also return values that are not positive integers. For example, consider the problem:
$$\log_{10}(0.1)$$

Which is equivalent to the exponential problem:
$$10^x = 0.1.$$

Since
$$0.1 = \frac{1}{10} = 10^{-1}$$

Then,
$$10^x = 0.1$$
$$10^x = 10^{-1}.$$

The bases are the same, but to be equal the exponents must also be the same. Equating the exponents, we see that $x = -1$.

In the last example we took the logarithm of 0.1. This is a number between 0 and 1. In fact, whenever the number we are taking the logarithm of is between 0 and 1, the output of the logarithmic function is always negative. This is because the only way we can obtain a value less that one by raising a base that is greater than one to a power, is if that exponent is negative. Consider the following:

Example 3: Find $\log_2(.125)$.

Solution: We must solve the related problem::
$$2^x = .125,$$

Rewrite the right hand side as a power of 2, i.e.
$$0.125 = \frac{1}{8} = \frac{1}{2^3} = 2^{-3}$$

Equating both sides we have that:

$$2^x = 2^{-3}.$$

Thus, $x = -3$.

It is not always possible to evaluate logarithms by hand. Consider the example:

$$\log_{10}(50)$$

Here, we want to find out: 10 raised to what exponent (power) is equal to 50, i.e.,

$$10^x = 50$$

The solution of this problem does not give us an easy integer solution. Here we have to do some thinking. We first recognize that $10^1=10$ and $10^2=100$. Since 50 is between these two numbers, it stands to reason that the value of *x* must be between 1 and 2.

We could start by guessing that $x = 1.5$ and compute $10^{1.5}$ on a calculator to obtain $10^{1.5}=31.6227...$ This value is too small, so we need to increase x. Let's increase it by 0.1 to 1.6 and keep changing it until our result is as close to the solution as we like, i.e. we increase the exponent and make a series of successive approximations until our result is sufficiently close to 50, i.e.,

$$10^{1.6} \approx 39.8107...$$
$$10^{1.7} \approx 50.1187...$$

Note: the value must be less than 1.7, but greater than 1.6 since the output is greater than 50. The result is relatively close, but if we want it to be more accurate than this, we have to choose a number less than 1.7 and continue until we get as close as we need to 50.

Continuing we have:
$$10^{1.65} \approx 44.6683...$$
$$10^{1.69} \approx 48.9778...$$
$$10^{1.6989} \approx 49.0019...$$
etc.

Luckily we don't have to do this all the time. Many calculators already have this capability. We only need to type in the expression correctly and let it do the work. Thus, on a calculator we simply type LOG(50) (the base 10 logarithm on a calculator is the LOG button) and out comes the value 1.69897000434…

This means that
$$10^{1.69897000434...} \approx 50$$

Example 4: Evaluate $\log_{10}(25)$

   Solutions: Using a calculator, we look for the LOG button and evaluate the expression as $\log_{10}(25) \approx 1.3979...$

Domain and Range of Logarithms:

Looking carefully at our definition for logarithms we can see that logarithms are not defined for negative numbers. In fact, logarithms are not even defined for 0. This is because there is no real number "x" such that $a^x=0$.

Logarithms can only be computed if the input is positive, or we say that the **domain** of a logarithm is all real numbers greater than zero, i.e., $x > 0$.

The output of a logarithm can be any real number, thus, the **range** of a logarithm is said to be all real numbers, i.e., $x \in \mathbb{R}$.

Example 5: Find $\log_4(-5)$.

   Solution: $\log_4(-5)$ is undefined, since there is no solution to the problem: $4^x = -5$. It is not possible to raise 4 to a power and get a negative number. We say that -5 is not in the domain of the logarithm.

Example 6: Find $\log_3(0)$.

   Solution: $\log_3(0)$ is undefined, since there is no solution to the problem: $3^x = 0$. It is not possible to raise 3 to a power and get zero. We say that 0 is not in the domain of the logarithm.

The Natural Base "e":

Previously we defined the natural exponential function,

$$f(x) = e^x.$$

The base of this function is "e." The logarithm associated with this base is called the natural logarithm. The natural logarithm can be written either as:

$$\log_e(x) \text{ or } \ln(x)$$

Both of these representations mean the same thing.

Example 7: Evaluate ln(2).

  Solution: Using our calculator we find $\ln(2) = 0.693147L$

Example 8: Evaluate ln(1).

  Solution: Using our calculator we find $\ln(1) = 0$. We also realized that we are solving the equivalent problem $e^x = 1$ and this is only true for $x = 0$.

Applications of Logarithms:

One of **Charles F. Richter**'s most valuable contributions was to recognize that the **seismic waves** radiated by all earthquakes can provide a good estimate of their magnitude. He collected the recordings of seismic waves from a large number of earthquakes, and developed a calibrated system of measuring them for magnitude.

Richter showed that the larger the energy of the earthquake, the larger the **amplitude** of ground motion at a given distance. He calibrated his scale of magnitudes using measured maximum amplitudes of waves on seismometers. Although his work was originally calibrated only for these specific seismometers, and only for earthquakes in southern California, seismologists have developed scale factors to extend Richter's magnitude scale to many other types of measurements on all types of seismometers, all over the world.

The Richter scale is the way we measure the severity of earthquakes. The scale is built around the base 10 logarithmic function. The Richter equation is given by:

$$M = \log_{10}(A),$$

where A is the amplitude, in millimeters, measured directly from the photographic paper record of the seismometer. M is the magnitude of the earthquake (the number we hear on the news, when they talk about a magnitude 5.7 earthquake).

Example 9: What is the Richter value of an earthquake whose seismograph gave an amplitude measurement of 10 mm.

  Solution: We use Richter's equation and obtain: $M = \log_{10}(10) = 1$.

Example 10: What is the Richter value of an earthquake whose seismograph gave an amplitude measurement of 1,000 mm.

Solution: We use Richter's equation and obtain: $M = \log_{10}(1,000) = 3$.

Notice that although the earthquake in example 10 was 100 times more powerful than the earthquake in example 9, the Richter value was only larger by two. Since the Richter scale is a logarithm we are actually measuring the exponent of the exponential function:

$$A = 10^M$$

Whenever, the Richter value increases by one, the strength of the earthquake has actually increased by a factor of 10! Thus, a magnitude 2 earthquake is 10 times greater than a magnitude 1, and a magnitude 9 earthquake is really 100 million times (10^8) more powerful than a magnitude 1 earthquake.

Another area in which logarithms are useful is in keeping track of the time required to perform searches on the Internet. A variety of methods have been used to search the Internet, but some types of searching methods take longer than others. Obviously, we would like the fastest method for searching otherwise we have to wait a while for the computer to return our results.

The most basic type of search is called a **linear search**. We illustrate this using an example. Suppose we wish to look up a person's name in the phone book. Let's say we are looking for the name "Ann Jones." In a linear search we simply open the phone book at page one and read each name until we find the name Ann Jones. This is not a very effective way of searching, especially if the list we are searching through is quite large.

A **binary search** is a more effective technique and we can describe it using the same example. We start by opening up the phone book approximately half way and look at the name at the top of the page. What does it say? Probably a name that begins with an 'L' or some letter near this. Now you think to yourself, does Jones come before or after this in the phonebook? Before, right? So you can ignore the entire last half of the phone book. Now open the remaining half about half way. You're probably somewhere near the 'E's. Does Jones come before or after 'E' in the phonebook? After, so you can ignore the beginning half. Continue doing this until you find the name for which you're looking. This is how a binary search is performed. We should note that a binary search only works for an ordered set of items (like an alphabetized list in a phone book).

A binary search is much faster than linear search for most data sets. If you look at each item in order in a linear search, you may have to look at every item in the data set before you find the one you are looking for. If a list contains n things, then a linear search can takes n operations to complete.

With the binary search, you eliminate half of the data with each decision. If there are n items in the list, then after the first decision you eliminate n/2 of them. After the second decision you've eliminated 3n/4 of them. After the third decision you've eliminated 7n/8 of them, etc. It turns out that this is equivalent to performing $\log_2(n)$ search operations to search the entire list with a binary search, instead of n using a linear search.

Example 11: A list contains 1,000,000 entries. How many operations will a linear and a binary search take to go through the entire list?

Solution: A linear search takes 1,000,000 operations.

A binary search takes $\log_2(1,000,000)$ operations. We cannot evaluate this directly using a calculator because we do not have a $\log_2$ button. Instead we use the formula: $\log_2(1,000,000) = \dfrac{\ln(1,000,000)}{\ln(2)} = \dfrac{\text{LOG}(1,000,000)}{\text{LOG}(2)}$ *. We can evaluate the right hand side of the equation using our computer or calculator and obtain: $\log_2(1,000,000) \approx 19.93$

* Here we use the BASIC representation for the natural logarithm, i.e. ln(x) = LOG(x).

BASIC Form of the Logarithmic Function:

The logarithmic function in BASIC only uses base $e \approx 2.718281828...$

LOG(x): Returns the value of ln(x) or $\log_e(x)$, the natural logarithm of x.

BASIC and mathematical notation: $\text{LOG}(x) = \log_e(x) = \ln(x)$
Domain: $x > 0$
Range: $\text{LOG}(x) \in \mathbb{R}$

This function is the inverse of EXP(x).

Examples:
1. LOG(-1) gives an error
2. LOG(0) gives an error
3. LOG(1)=0
4. LOG(2.718281828)=1

Note: The only base built into BASIC is *e*. However, it is possible to find logarithms in any positive base, subject to the limitations of memory. This, however, is beyond the scope of what we want to talk about in this book. Whenever we need to compute a logarithm in a different base on the computer, we will give you the conversion formula.

Objective G: Properties of Logarithmic Functions (Optional)

One of the main uses for logarithms is in calculations. Logarithms turn multiplication into addition, division into subtraction and exponentiation into multiplication. Using these characteristics of logarithms it is possible to simplify complicated calculations. In the days before calculators, this was very important. Although logarithms are no longer used as frequently to simplify calculations, they are still extremely important functions that can be used to understand and model problems. We now look at the logarithmic properties that enabled these simplifying transformations.

When we introduced exponents in Chapter 1, we developed several different exponent rules (laws). The first four are rewritten here for convenience:

1. $b^n * b^m = b^{(n+m)}$
2. $\dfrac{b^n}{b^m} = b^{(n-m)}$
3. $(b^n)^m = b^{n*m}$
4. $b^0 = 1$

As we have already shown, logarithms are exponents, so it seems only natural that there are corresponding rules for logarithms.

In this section we will state the related rules for logarithms and also show how to use them.

The laws apply equally to all logarithms of any base. We will write the laws using the logarithm associated with the base e, i.e., $\ln_{10}$, which we write as ln. The laws, however, apply to all bases.

Addition of Logarithms: ln (a*b) = ln (a) + ln (b)

"The logarithm of a product equals the sum of the logarithms"

Examples:
1. ln(3*x) = ln(3) + ln(x)
2. $\ln(x^2 y^3) = \ln(x^2) + \ln(y^3)$

Logarithms turn multiplication into addition.

Subtraction of Logarithms: ln (a/b)= ln (a) – ln (b)

"The logarithm of a quotient equals the difference of the logarithms"

Examples:

1. $\ln\left(\dfrac{x}{3}\right) = \ln(x) - \ln(3)$
2. $\ln\left(\dfrac{7}{x}\right) = \ln(7) - \ln(x)$

Logarithms turn division into subtraction.

Multipes of Logarithms: ln (aⁿ) = n ln (a)

Examples:

1. $\ln(x^2) = 2\ln(x)$
2. $\ln(9) = \ln(3^2) = 2\ln(3)$

Logarithms turn exponentiation into multiplication.

Zero Exponent Logarithms: ln (1) = 0

Examples:

1. $\ln(10^0) = \ln(1) = 0$
2. $\ln(x^0) = \ln(1) = 0$, x>0

The logarithm of any expression raised to the zero power is zero. Thus, log(1) is always zero, no matter what the base is.

EXERCISES 4.2

Evaluate each of the following:
1. ABS(-5)
2. 67 MOD 8
3. 103 MOD 11
4. SQR(-36)
5. 39 MOD 6
6. 49 MOD 5
7. ABS(0)
8. 47 MOD 4

9. 28 MOD 3

10. SQR(24)

Write each equation in logarithmic form.

11. $2^3 = 8$

12. $2^6 = 64$

13. $3^2 = 9$

14. $2^{-3} = \dfrac{1}{8}$

15. $5^{-4} = \dfrac{1}{625}$

16. $81^{1/2} = 9$

Write each equation in exponential form.

17. $\log_2(8) = 3$

18. $\log_5(125) = 3$

19. $\log_{81}(9) = \dfrac{1}{2}$

20. $\ln(e^3) = 3$

21. $\log_{10}\left(\dfrac{1}{100}\right) = -2$

22. $\log_2\left(\dfrac{1}{64}\right) = -6$

23. $\log_3(x) = 3$

Evaluate the following.

24. $\log_2(32)$

25. $\log_{10}(1000)$

26. $\log_3(27)$

27. $\log_{10}\left(\dfrac{1}{100}\right)$

28. $\log_2\left(\dfrac{1}{16}\right)$

29. $\ln\left(\dfrac{1}{e}\right)$

30. $\ln(e^{-3})$

Evaluate using a calculator. Write your answer in floating point notation with 4 significant digits and rounding.

31. $\ln(6)$

32. $\log_{10}(43)$

33. $\log_{10}(1024)$

34. $3^{2.5}$

35. e^{-3}

36. 2^{-10}

Use the properties of logarithms to expand.

37. $\log_3(4*10)$

38. $\log_2[7(x+2)]$

39. $\ln\left(\dfrac{5}{19}\right)$

40. $\log_8\left(\dfrac{x}{x+1}\right)$

41. $\log_2(2^7)$

42. $\log_5(5^{-3})$

43. $\log_4\left[3^3(x-1)^2\right]$

44. $\log_2\sqrt{x^2+1}$

Use the properties of logarithms to write as a single logarithm.

45. $\log_3(4)+\log_3(2)$

46. $\log_2(5)+\log_2(x)$

47. $\log_{10}(25)-\log_{10}(5)$

48. $\log_3(28)-\log_3(7)$

49. $3\log_2(3)+2\log_2(x)$

50. $\dfrac{1}{2}\log_4(81)$

COMPUTER ACTIVITY

1. Write a BASIC program that will input the amplitude of the line measured in millimeters on a seismograph and out put the magnitude of the earthquake. You must use the formula:
$$M = \log_{10}(A);$$

   however, BASIC does not have the base 10 logarithm available directly. It only has the built in base "e." To get around this we use the conversion formula:

$$\log_{10}(A) = \frac{LOG(A)}{LOG(10)}.$$

   Thus, we use the formula:
$$M = \frac{LOG(A)}{LOG(10)}$$

   Run the program for the following amplitude values and print out the results:

   22mm, 145mm, 10,340mm, 1,950,000mm, and 9,999,999,999mm.

2. Write a BASIC program that will input the total number of items in a list to be searched (n) and output the maximum number of inquiries that must be made to the list to find the item of interest. Use the formulas for both a linear and a binary search.

   Linear search: Number of inquiries = n
   Binary search: Number of inquiries = $\log_2(n)$

   Again, we cannot implement this directly, since BASIC does not have the base 2 logarithm. Instead we must use the conversion:

$$\log_2(n) = \frac{LOG(n)}{LOG(2)}$$

   Run the program for the following list sizes and print out the results for both the linear and binary search:

   100, 21987, 295000, and 19000000000.

Section 4.3: Additional Built-in BASIC Functions (Optional)

BASIC has many built-in functions. We will present some additional examples here. First, we must keep in mind that BASIC performs operations on real numbers (R) and subsets of real numbers, such as integers (Z) and rational numbers (Q).

Some additional built-in functions that are useful are:

SGN(x): Returns the sign of a numeric expression. The value 1 will denote a positive number, 0 will denote a 0 and –1 will signify a negative number.

BASIC and mathematical notation: $\text{SGN}(x) = \begin{cases} 1 \text{ if } x > 0 \\ 0 \text{ if } x = 0 \\ -1 \text{ if } x < 0 \end{cases}$

Domain: $x \in R$
Range: $\text{SGN}(x) \in \{-1, 0, 1\}$

This is known as the *signum* function, from the Latin for "sign". We call this the *signum* function so as not to confuse it with the sine (pronounced "sign") function in trigonometry.

Examples:
1. SGN(123)=1
2. SGN(-14.56)=-1
3. SGN(0)=0
4. SGN(33-48)=-1

INT(x): Returns the greatest integer less than or equal to a numeric expression.

BASIC and mathematical notation: $\text{INT}(x) = [\![x]\!]$ the greatest integer $\leq x$
Domain: $x \in R$
Range: $\text{INT}(x) \in Z$

This is also known as the "floor" function.

This function neither rounds nor truncates the argument, but finds the greatest integer less than or equal to the argument.

Examples:
1. INT(16)=16
2. INT(1.414214)=1

3. INT(2.999)=2
4. INT(-31.6)=-32
5. INT(-7.1)=-8

FIX(x): Returns the integer part a numeric expression.

BASIC and mathematical notation: $FIX(x) = \begin{cases} INT(x) \text{ if } x \geq 0 \\ -(INT(-x)) \text{ if } x < 0 \end{cases}$

Domain: $x \in R$
Range: $FIX(x) \in Z$

This function truncates the argument by ignoring the fractional (decimal) part of the number.

The above definition, although convoluted, states that in the case of a negative argument, we negate the argument, find its floor, and negate it again to get the correct negative integer part.

Examples:
1. FIX(16)=16
2. FIX(1.414214)=1
3. FIX(2.999)=2
4. FIX(-31.6)=-31
5. FIX(-7.1)=-7

LEN(A$): Returns the number of characters in string A$.

BASIC notation: LEN(A$)
Domain: A$ is a character string
Range: LEN(A$) ∈ N

This function takes a character string and computes the number of characters in the string.

Examples:
1. LEN("cat")=3
2. LEN(1.414214)=Undefined, input must be a string
3. LEN("01234")=5
4. LEN("good luck")=9, the space counts as a character
5. If A$="MATH", then LEN(A$) = 4

COMPUTER ACTIVITY

1. Write a BASIC program that inputs a number, x, and outputs both the original number (x) and the SGN(x).

    Write the program output for the following inputs: 0, -13, 451, -389, 293.

2. Write a BASIC program that inputs a number, x, and outputs both the original number (x) and the INT(x).

    Write the program output for the following inputs: 0, -13.25, 4.59, -38.9, 29.3.

3. Write a BASIC program that inputs a number, x, and outputs both the original number (x) and the FIX(x).

    Write the program output for the following inputs: 0, -23.75, 0.251, -3.4798, 22.4.

Section 4.4: Subroutines and User Defined Functions

Large programs can get extremely complicated to write and to follow. It is often necessary to divide the program up into more workable sub-parts. This helps make the program easier to read and ultimately reduce programming errors and debugging time. BASIC provides ways to organize and sub-divide a program into workable sub-parts. In BASIC we can either define a structure called a **subroutine** or create a user defined **function**.

Objective A: Subroutines

A *subroutine* (*also called subprogram*) is a sequence of program statements that performs one or more related tasks. The main program can access subroutines by using a **CALL** statement. The subroutine is written as a separate part of the program.

The following is a simple example of a subroutine. Here we wish to create a subroutine that simply prints a message on the output screen.

Example 1: Type in and run this program.
```
REM display the product of two numbers
CLS
CALL Explain
END

SUB Explain()
    PRINT "This program will multiply two numbers."
END SUB
```

In this example we point out two things. First the Main program is written as usual with an **END** statement at the bottom of the program. The subroutine called Explain is created outside (beneath) the Main program with the **SUB** command and is completed with the **END SUB** command. To identify which subroutine we are referring to, since a program can contain more than one subroutine, the name of the subroutine being called is placed next to both the **CALL** and **SUB** commands in the program.

It is also possible to pass data to and from a subroutine. In the previous example, we did not pass any data, but notice that we had to include a pair of opening and closing parentheses next to the subroutine name in the **SUB** command line. This is required in BASIC otherwise the program will not run.

As another example, let's modify the first program above to include another subroutine called Multiply. The Multiply subroutine will take two input values, and print an output that is simply the product of the two numbers.

Example 2: Type in and run this program.

```
REM display the product of two numbers
CLS
CALL Explain
PRINT
x = 2
y = 3
CALL Multiply(x, y)
END

SUB Multiply(a, b)
   REM display numbers and their product
   PRINT "The product of"; a; "and"; b; "is"; a * b

END SUB

SUB Explain()
    PRINT "This program will multiply two numbers."
END SUB
```

The variables in the parenthesis of subroutine name are called **parameters**. Notice that in this example we had to pass both the values we wanted to find the product of. Also note that we can give the variables different names in the CALL and SUB command lines and the program automatically takes care of equating the correct variables. The number of values being passed in the CALL statement must always equal the number of values in the SUB statement. In the subroutine Multiply we passed 2 values (x and y) in the CALL statement, so we must have 2 values (a and b) in the SUB statement.

Can you modify it so that it can multiply other pairs of numbers? Try for example: 4 and 5, 10 and 12, and 35 and 6. What will your output look like?

Example 3: Type in and run this program using a tax rate of 6.5% and 8% and at a sweater cost of $45 and $65.

```
REM  This program calculates the total amount paid
REM  for an item including tax.
INPUT  "Enter the tax rate in decimal"; tax
CALL   InputCost ("sweater", cost)
CALL   CalculateItem(cost, tax, total)
CALL   DisplayTotal  (total)
END
```

```
        SUB  InputCost (item$, cost)
              PRINT   "What is the price of a "; item$;
              INPUT   cost
        END SUB

        SUB CalculateItem(cost, tax, total)
              total = cost *(1 + tax )
        END SUB

        SUB DisplayTotal  (total)
              PRINT  " The total amount paid for the
              item is";
              PRINT USING "$###.##"; total
        END SUB
```

OUTPUT:

```
```

Objective B: User Defined Functions

A *function* can be a built-in function or a user-defined function. We've already introduced a number of built-in functions in BASIC, such as **INT** (to find the greatest integer less than or equal to the number in the parenthesis), **LEN** (gives the number of characters in a given string), and **SQR** (gives the square root of a given number).

User-defined functions are similar to subroutines and are used the same way as the built-in functions. They reside outside the main body of the program and are defined by function blocks of the form

```
    Function    FunctionName( var1, var2, . . . .)
                    .
                    .
                    .
        FunctionName = expression
    END FUNCTION
```

The variables in the parenthesis of FunctionName are called **parameters**.

Subroutines and functions are called **procedures.** They can be used in other programs; hence, they eliminate repetitive codes.

Example 1: What will the output of the following program be if you input the following pairs of triangle side lengths: (3,4), (4,8), and (9,11)

In the following example the built-in function **SQR** as well as a user-defined function Hypotenuse(A,B) are used.

```
REM    This program will calculate the length of the
       hypotenuse of a right triangle.
CLS
INPUT  "Enter the lengths of the two sides of the
       right triangle: ", A, B
PRINT  " The hypotenuse of the right triangle is
       "; Hypotenuse(A,B)
END

FUNCTION   Hypotenuse(A,B)
           Hypotenuse = SQR(A^2+B^2)
END FUNCTION
```

In the above program, instead of a subroutine, we used a function. This function is not a built-in function such as **SQR (square root function)**. This function is actually executing a formula defined within the subroutine.

What is the formula that we used in this example? _____

What is the output of this program?

OUTPUT:

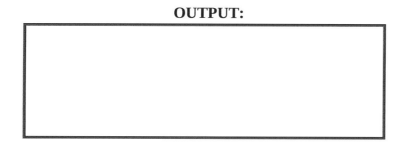

COMPUTER ACTIVITY

Student Name_____ Section_____ Date_____

Finding Roots of a Quadratic Equation

A quadratic equation is of the form $ax^2 + bx + c = 0$ where **a, b,** and **c** are the coefficients of the variables and constant. To find the roots or to solve this quadratic equation, we can use the quadratic formula;

$$x = \frac{-b \pm \sqrt{b^2 - 4ac}}{2a}$$

The discriminant $b^2 - 4ac$ determines whether the roots are real or complex (imaginary). If $b^2 - 4ac < 0$, then there are two complex solutions. If $b^2 - 4ac = 0$, then there is only one solution which has multiplicity of 2. If $b^2 - 4ac > 0$, then there are two real solutions.

Use the quadratic formula to solve the following quadratic equations:

Activity 1. $2x^2 - 3x + 1 = 0$

Activity 2. $x^2 + 3x + 12 = 0$

Activity 3. $x^2 - 4x + 4 = 0$

BASIC Programming

Use the quadratic equations above to write a program with an **INPUT** statement to enter the coefficients of the equation and use the discriminant as a user-defined function to determine what type of solutions each has.

Also, use the quadratic formula as either a function or a subroutine to solve the equation.

What is displayed when the solutions are complex?

COMPUTER ACTIVITY

Student Name_____ Section_____ Date_____

Which Day of the Week does January 1 of a given year occur?
(use of functions MOD and INT)

The following formula can be used to find the day of the week on which a given year begins. (Formula appears in *Mathematical Ideas*, by Miller, Heeren & Hornsbey, 9th edition, Addison Wesley Publishing)

$$a = y + [\![(y-1)/4]\!] - [\![(y-1)/100]\!] + [\![(y-1)/400]\!],$$

where *y* represents the given year in 4 digits.

After finding *a*, calculate the number $b = a \bmod 7$. Then *b* gives the day of the week of January 1, with $b = 0$ representing Sunday, $b = 1$ Monday, $b = 2$ Tuesday and so on. Find the day of the week on which January 1 would occur in the following years.

Activity 1. 1812

Activity 2. 2001

Activity 3. 2020

Activity 4. 1988

BASIC Programming

Write a program requesting the user to **INPUT** the year and have the program calculate the values of both *a* and *b*. Your output should include the values of *a* and *b*. It should also include a sentence that states:

January 1 of the year **xxxx** is on a **(day of the week)**.

COMPUTER ACTIVITY

Student Name_____ Section_____ Date_____

Which Day of the Week does a given date occur?
(use of functions MOD and INT)

The following formula can be used to find the day of the week given any date after the year 1582. This formula is called the Zeller's Congruence.

Let D = day of the month, M = month (13 for January and 14 for February, 3 for March, 4 for April and so on) and

Y = year in four digits. For January and February, subtract 1 from the given year.

$$a = D + \left\lfloor \frac{26*(M+1)}{10} \right\rfloor + \left\lfloor \frac{125*Y}{100} \right\rfloor - \left\lfloor \frac{Y}{100} \right\rfloor + \left\lfloor \frac{Y}{400} \right\rfloor - 1$$

where $\lfloor x \rfloor$ represents the greatest integer less than or equal to x.

After finding a, calculate $b = a \bmod 7$. Then b gives the day of the week with $b = 0$ representing Sunday, $b = 1$ Monday, $b = 2$ Tuesday and so on. Then try the formula for the following dates.

Activity 1. 1776   *Activity 2.* 2004   *Activity 3.* 2035   *Activity 4.* 1984

BASIC Programming

Write a program requesting the user to **INPUT** the year and the number and day of the month. Use the formula above to calculate a, then calculate b using a MOD 7. Treat January as the 13th month and February as the 14th month of the previous year. The remainder when a is divided by 7 (that is, a MOD 7) is the day of the week, with 0 assigned to Sunday, 1 assigned to Monday and so on.

Then for the output, print the sentence: Date is a day of the week.
NOTE: You will have to use decision structures (such as IF-THEN) in this program.

Section 4.5: Repetition Structures

Suppose we want to keep running a program for a number of different inputs, but do not want to keep re-starting the program. We can do this using the *repetition structure* or *loops*. Refer to the manual in Appendix A for more information on repetition structures.

Objective A: FOR-NEXT Loop

The first type of looping structure we consider is called the FOR-NEXT loop. In this structure looping variables identify the starting (initial), stopping (limit) and incremental value of a counting variable. We declare the starting, stopping and increment using a **FOR** command and then the increment is performed using a **NEXT** command. All the code between the **FOR** and **NEXT** statements are executed until the looping variable is greater than the stopping (limit) value if the increment is positive and less than the stopping value if the increment is negative.

The flow chart below indicates the logical flow of this looping structure for a positive increment.

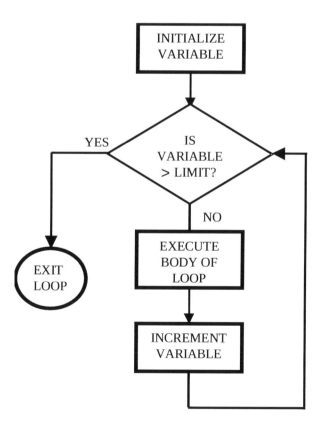

The loop is set up and executed using a FOR and a NEXT command.

FOR – NEXT statement: Executes the code between the **FOR** and **NEXT** statements a fixed number of times. If STEP is omitted, then the increment is 1.

FOR *variable = initial* **TO** *limit* **[STEP** *increment***]**

$$\underline{} \underline{} \underline{}$$
$$\underline{} \underline{} \underline{}$$
$$\underline{} \underline{} \underline{}$$

NEXT *variable*

To illustrate the FOR-NEXT statement, suppose we want to calculate the salary of 10 employees, given the hours and their respective hourly rates, we use the FOR-NEXT loop which states exactly how many times the sequence is to be executed:

Example 1:

```
REM    This program will calculate weekly salary
REM    for 10 employees
REM    Input the numbers of hours worked and the hourly rate
CLS
FOR    I = 1 to 10
INPUT    "Enter the number of hours worked ";hours
INPUT    "Enter the hourly rate"; rate

SELECT    CASE hours
          CASE  IS >=50
                salary = 40*rate + 2*rate*(hours - 40)

          CASE  IS  > 40

                salary = 40*rate + 1.5*rate*(hours - 40)
          CASE  IS  < = 40
                salary = hours * rate
END SELECT
PRINT    "hours worked: ";hours, "hourly rate: ";rate, "weekly
     salary:"; salary
NEXT    I
END
```

Objective B: DO-WHILE Loop

Another repetition structure that could be used is called the DO-While or the ***pre-test loop***. In this type of loop, the exit test, the statement that contains the exit condition, is stated at the ***top*** of the loop. If the exit condition is met, then the loop stops and the rest of the program will be executed.

The flow chart below indicates the logical flow of this looping structure.

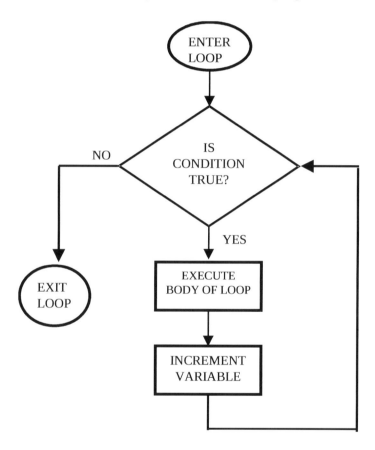

The loop is set up and executed using a DO WHILE and a LOOP command as illustrated below.

DO WHILE *Loop*: The condition is evaluated. If true, the body of the loop is executed, and the program goes back to the top once it reaches LOOP; if false, then the program branches to the statement after LOOP.

DO WHILE *condition*

```
— — —
— — —
— — —
```
LOOP

To illustrate the DO-WHILE statement, we use the same problem in example 1 above, only now we use the DO-WHILE structure.

Example 2:

```
REM    This program will calculate weekly salary for
REM  several employees using pre-test loop.
REM    Input the numbers of hours worked and the hourly rate
CLS
counter =0    ' this is to initialize the counter
DO WHILE  counter <10 ' For 10 employees
INPUT    "Enter the number of hours worked ";hours
INPUT    "Enter the hourly rate"; rate
SELECT   CASE hours
          CASE  IS  >=50
                   salary = 40*rate + 2*rate*(hours - 40)

          CASE   IS  > 40
                   salary = 40*rate + 1.5*rate*(hours - 40)
          CASE   IS  < = 40
                   salary = hours * rate
END SELECT
PRINT    "hours worked: ";hours, "hourly rate: ";rate, "weekly
     salary: ";salary
Counter=counter + 1
LOOP
END
```

Objective C: DO-UNTIL Loop

A third way of doing repetition structure is the use of *post-test loop*. That is the exit condition appears at the **bottom** of the loop.

The flow chart below indicates the logical flow of this looping structure.

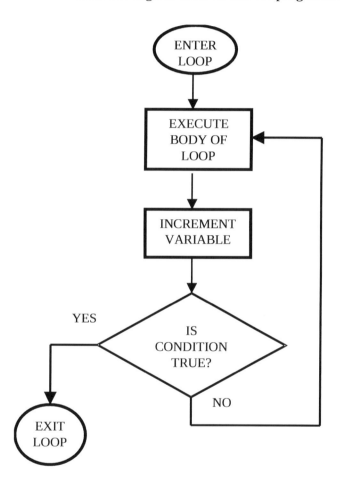

DO ...UNTIL *Loop*: The body of the loop is executed, the condition is evaluated. If true, the program branches to the statement after LOOP UNTIL; if false, then the program goes back to the top of the loop.

DO _ _ _
       _ _ _
       _ _ _

LOOP UNTIL *condition*

Using the same example we demonstrate the DO-UNTIL or the *post-test loop*:

Example 3:

```
REM   This program will calculate weekly salary for
REM   several employees using Post-test loop.
REM   Input the numbers of hours worked and the hourly rate
counter = 0    'initializing the counter to 0
DO
        INPUT   "Enter the number of hours worked"; hours
        INPUT   "Enter the hourly rate"; rate

     SELECT   CASE hours
         CASE  IS >=50
                 salary = 40*rate + 2*rate*(hours - 40)

         CASE  IS  > 40

                 salary = 40*rate + 1.5*rate*(hours - 40)
         CASE  IS  < = 40
                 salary = hours * rate
      END SELECT

     PRINT    "hours worked: ";hours, "hourly rate: ";rate,
    "weekly salary: "; salary
Counter = counter +1
LOOP UNTIL counter = 10 ' For 10 employees
END
```

For examples 1-3 above, try the following hours and rate. Compare the output for each program:

1. 45 hours and $8.00 hourly
2. 38 hours and $9.00 hourly
3. 62 hours and $12.00 hourly

More Examples of Repetition Structures

1. What is the output of the following program for num = 8, 13, and 21?

   ```
   'Example of For-Next loop
   CLS
   INPUT "ENTER A NUMBER"; num
   FOR I = 1 TO num STEP 2
     PRINT I
   NEXT I
   END
   ```

   OUTPUT:

2. What is the output of the following program?

   ```
   'Example of Query-controlled loop
   CLS
   DO
    PRINT "Enter employee name, hours worked, and rate"
     PRINT "         (separated by commas)"
     PRINT
     INPUT "--->", Employee$, Hours, Rate
     LET Salary = Hours * Rate
     PRINT USING "  Gross Salary...$$####,.##"; Salary
     PRINT
     INPUT "Process another employee?(Y/N)", Response$
     PRINT
   LOOP UNTIL Response$ <> "Y"
   ```

   OUTPUT:

3. What is the output of the following program?

```
'Example of nested loop
CLS
FOR Count = 1 TO 4
  FOR Asterisk = 1 TO Count
      PRINT "*";
  NEXT Asterisk

  PRINT TAB(14);
  FOR number = Count TO 4
      PRINT "#";
  NEXT number
  PRINT
NEXT Count
END
```

OUTPUT:

```
*            ####
**           ###
***          ##
****         #
```

4. What is the output of the following program?

```
REM  An example of nested loop
FOR I = 1 TO 3
     FOR J = 10 TO 1 STEP -2
       PRINT J,
     NEXT J
   PRINT
NEXT I
```

OUTPUT:

```
10    8     6     4     2
10    8     6     4     2
10    8     6     4     2
```

EXERCISES 4.5

1. Explain the difference between a FOR-NEXT and a DO-WHILE loop.
2. Explain the difference between a FOR-NEXT and a DO-UNTIL loop.
3. Explain the difference between a DO-WHILE and a DO-UNTIL loop.

COMPUTER ACTIVITY

Student Name_____ Section_____ Date_____

Celsius versus Fahrenheit
(repetition)

Temperatures can be measured in Celsius or Fahrenheit. The freezing point is 0° in Celsius and 32° in Fahrenheit while the boiling point is 100° in Celsius and 212° in Fahrenheit, as discussed in Section 4.2.

To convert a temperature that is given in Celsius to Fahrenheit, we use the formula
$F = \frac{9}{5}C + 32$, where **F** is degrees in Fahrenheit and **C** is degrees in Celsius.

To convert temperature that is given in Fahrenheit to Celsius, we use the formula
$C = \frac{5}{9}(F - 32)$, where **C** is degrees in Celsius and **F** is degrees in Fahrenheit.

If the temperature given is in Celsius, then convert it to Fahrenheit. If the temperature is given is Fahrenheit, then convert it to Celsius.

Activity 1. 84° F   *Activity 2.* 37° C

BASIC Programming

Write a program that will ask the user whether the temperature is given in Celsius or Fahrenheit. If the temperature given is in Fahrenheit, then convert it to Celsius, and vice – versa. Your output should display the temperature entered and its conversion. Include a counter that will count how many temperatures were converted. Place a limit (say 10) to the number of conversion to be done.

Test your program with the following temperatures: 68° **F** , 120° **C**, -20° **F** , 37° **F**, 88° **F** , 22° **C**, 50° **C** , 189° **F**, 45° **F** , and 55° **C**

Your output should look like the following:

68 degrees Fahrenheit is equal to _____ degrees Celsius.
120 degrees Celsius is equal to _____ degrees Fahrenheit. Etc.....

Section 4.6 Recursion and Recursive Functions

Recursion is the terminology we use when a function calls itself. It is similar to a loop because it repeats the same code. The difference is that the recursive function automatically passes the looping variable back into itself, without a loop ever being set up. Many, but not all, programming languages allow recursion because it can simplify some tasks, and it is often more elegant than a loop.

Factorials and Recursion

The easiest way to introduce recursion is through the mathematical function called a factorial. A factorial is a special mathematical function that takes a number and then successively multiplies it by one less than that number until the value of 1 is reached. A factorial is written using the exclamation point, "!" placed after the number. For example 5!, read "five factorial," is simply the successive product of descending counting numbers, i.e. 5*4*3*2*1=120.

Examples of factorials are:

$$1! = 1$$
$$2! = 2*1 = 2$$
$$3! = 3*2*1 = 6$$
$$4! = 4*3*2*1 = 24$$
$$5! = 5*4*3*2*1 = 120$$
$$6! = 6*5*4*3*2*1 = 720$$
$$\vdots$$
$$\text{etc.}$$

In general we write:

$$n! = n*(n-1)*(n-2)*(n-3)* \ldots *2*1$$

We should also note that $0! = 1$.

An interesting and sometimes useful property of a factorial is illustrated in the following:

$$5! = 5*4!$$
$$6! = 6*5!$$
$$\vdots$$
$$n! = n*(n-1)!$$

This is the recursive form of the factorial function, i.e.

$$n! = n*(n-1)!$$

Examples: Evaluate each of the following:
1. 5!
2. 8!
3. 2!

Solutions:
1. 5! = 5*4*3*2*1 = 120
2. 8! = 8*7*6*5*4*3*2*1 = 40,320
3. 2! = 2*1 = 2

We can calculate a factorial recursively. The program below represents the recursive solution for the factorial problem. Type in and then run the following program for different inputs:

```
CLS
INPUT "Input the number you want to find the factorial of";N
PRINT "The solution is ";N;"! = "; fact(N)
END

FUNCTION fact (N)
   IF N > 1 THEN
      fact = N * fact(N - 1)
   ELSE
      fact =1
   END IF
END FUNCTION
```

In this program we defined a function called "fact" that simply takes the input and multiples it by the factorial of one less than the input. Notice too that the function calls itself recursively with the line: `fact = N * fact(N - 1)`.

It is also possible to compute this factorial using a simple loop in BASIC. The following program will compute the factorial of a number n using a FOR-NEXT loop:

```
CLS
INPUT "Input the number you want to find the factorial of";N
N1=N
```

```
FOR (I=1 TO N-2)
  N1=N1*(N-I)
NEXT I
IF (N=0) THEN N1=1
PRINT "The solution is ", N, "! = ", N1
END
```

Type this program in and run it and observe the output for several different factorials.

We have shown that the factorial example can be solved either by iteration (looping) or recursion. There are, however, some problems that can only be solved through recursion. An example would be to have a computer display all folders and sub-folders in a file structure. This we'll call the folder display problem.

In this example we do not know ahead of time how many folders and subfolders there are in our file system. Thus, we cannot set up a loop from the start. We simply have to open each folder, display its name, and obtain a list of its contents and then open up every folder within this folder and display their contents and continue doing this until we have displayed all folders.

This is where a recursive function can be used. We could simply define a function whose purpose is to display the name of the current folder and obtain a list of all its subfolders and repeat the function for each of the subfolders.

In the factorial problem, recursion is not necessary because we know beforehand that the loop will be executed until the value of the number we are decrementing and multiplying by is equal to 1. However, in the folder display example, we never know how many subfolders are present in a folder and how many iterations will be required. For this purpose, the only solution is to call the function recursively to display information about a folder and its subfolders.

A word of warning, you must use the recursion technique carefully. If you fail to implement it appropriately, you can end up with an application that never ends.

Examples: *Evaluate the following:*
   4. 5!
   5. 8!
   6. 2!

   Solutions:
   4. 5! = 5*4*3*2*1 = 120

5. $8! = 8*7*6*5*4*3*2*1 = 40,320$
6. $2! = 2*1 = 2$

Fibonacci Numbers and Recursion

In 1202 Leonardo Fibonacci of Pisa introduced a problem that asked the question; how many pairs of rabbits can be generated from a single pair, if each month each mature pair gives birth to a new pair, which only after two months also becomes productive, i.e.,

Pairs	Month	Reason
1	1	New starting pair
1	2	First pair not mature to give birth until two months old
2	3	First pair produced a second pair: (1 + 1).
3	4	First pair, plus a new pair from first pair, plus the second pair: (2 + 1).
5	5	First pair, plus a new pair from first pair, plus the second pair, plus a new pair from the second pair, plus the third pair(from pair 1 above).(3 + 2).
8	6	First pair, plus a new pair from first pair, plus the second pair, plus a new pair from the second pair, plus the third pair, plus a new pair from the third pair, plus the fourth pair (from pair 1 above), plus the fifth pair (from pair two above) (5 + 3).

Can you see a pattern emerging?

To get the next number in this sequence you add together the two previous numbers. Thus, the seventh number in the sequence is obtained by adding 8 and 5 together, or $8 + 5 = 13$, etc.

This sequence of numbers: 1, 1, 2, 3, 5, 8, 13, 21, … is named after Fibonacci and is called a **Fibonacci Sequence** (or Fibonacci numbers).

The nth term (number) in a Fibonacci sequence is denoted by, $F(n)$, i.e. $F(1) = 1$, $F(5) = 5$, and $F(6) = 8$.

Examples: *Find the indicated Fibonacci number*
1. $F(2)$
2. $F(10)$
3. $F(-1)$

Solution:
1. $F(2) = 1$, since the sequence begins with 1, 1.

2. F(10) = 55, since the sequence is 1, 1, 2, 3, 5, 8, 13, 21, 34, 55, and F(10) = F(9) + F(8) = 34 + 21 = 55
3. F(-1) is undefined. The Fibonacci number cannot be computed for negative numbers

The recursive form of the Fibonacci Sequence is:

$$F(N) = F(N-1) + F(N-2)$$

The Pyramid Problem and Recursion

Consider a three-sided pyramid (sometimes called a tetrahedron). To make the pyramid, we start with a triangular arrangement of tennis balls for the bottom layer, e.g.,

Next stack up tennis balls a layer at a time.

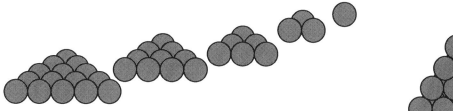
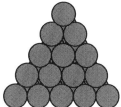

If the base of the pyramid has N balls along a side, how many balls does the entire pyramid have in it?

We call this a **pyramidal number** and write it as Pyramid(N).

Here is a chart that shows the number of balls inside a pyramid (pyramidal number) whose base is N balls on a side.

N	1	2	3	4	5	6	7
Pyramid(N)	1	4	10	20	35	56	84

In filling out the chart you might have noticed that it would be useful to know how many balls are in each layer of the pyramid. Of course, the number of the balls in the pyramid is the sum of the number of balls in each layer. But, the number of balls in layer N is the number of balls in a triangle with a base of N balls. The number of balls in a triangular layer is called the **triangular number** and is written as Triangle(N).

Thus, the number of balls in a triangle and a pyramid as a function of the number of layers N is given as:

N	1	2	3	4	5	6	7
Triangle(N)	1	3	6	10	15	21	28
Pyramid(N)	1	4	10	20	35	56	84

Looking at the chart, we see the following patterns.

For the triangular numbers we have that:

$$\text{Triangle}(2) = 2 + \text{Triangle}(1)$$
$$\text{Triangle}(3) = 3 + \text{Triangle}(2)$$
$$\text{Triangle}(4) = 4 + \text{Triangle}(3)$$
$$\vdots$$
$$\text{Triangle}(N) = N + \text{Triangle}(N-1)$$

Then looking along the diagonal of the chart above, we see the following pattern for the pyramidal numbers:

$$\text{Pyramid}(2) = \text{Pyramid}(1) + \text{Triangle}(2)$$
$$\text{Pyramid}(3) = \text{Pyramid}(2) + \text{Triangle}(3)$$
$$\text{Pyramid}(4) = \text{Pyramid}(3) + \text{Triangle}(4)$$
$$\vdots$$
$$\text{Pyramid}(N) = \text{Pyramid}(N-1) + \text{Triangle}(N)$$

This is the recursive definition for a pyramidal number.

EXERCISES 4.6

1. What is recursion?

2. Explain why the factorial function can be implemented on a computer with or without recursion.

3. *Evaluate each of the following:*
    a. 4!
    b. 7!
    c. 12!
    d. F(11)
    e. F(0)
    f. F(14)

COMPUTER ACTIVITY

Student Name_____ Section_____ Date_____

1. Write a program to implement the Fibonacci Sequence defined recursively as:

   $$F(N) = F(N-1) + F(N-2)$$

   Have the program print out the first 100 Fibonacci numbers.

2. Write a program to implement the pyramid problem defined recursively as:

   $$Pyramid(N) = Pyramid(N-1) + Triangle(N)$$

   Where,

   $$Triangle(N) = N + Triangle(N-1)$$

   Have the program print out the pyramidal number for the first 15 pyramids.

Chapter 5

Subscripts, Vectors, Matrices, and Arrays

One of the most important capabilities of a computer is its ability to work with and process large sets of data. Suppose a company wants to sort the salary of its 1000 employees from lowest to highest. Furthermore, the company may want to find out how many of its 1000 employees earn $50,000 and make a list of those employees' names. The company may even want to calculate the average salary of its employees.

To try to do what this company wants is labor intensive as well as time consuming. In this chapter, we will show how to organize different types of data, both numeric and character string, so that we can perform certain tasks or manipulate the data. We'll introduce the concepts of matrices and arrays, one-dimensional as well as two- dimensional. We'll also show some operations on matrices and how these are performed in BASIC programming.

After completing this chapter you should understand the basic properties of vectors (one-dimensional arrays) and matrices (two-dimensional arrays) and be able to:

- identify the elements of an array,
- add and subtract vectors,
- multiply vectors by scalar quantities,
- compute the inner product of two vectors,
- identify the elements and order of a matrix,
- add, subtract, and multiply matrices,
- declare and work with arrays on a computer, and
- work with the summation and product operators.

Section 5.1: One-Dimensional Arrays: Vectors

The real strength of a computer is that it can process large amounts of data very rapidly. Problems arise, however, in organizing all this data as we are processing it. If we were to use the tools we have learned up until now, we would assign each piece of data to a new variable. This would require spending a great deal of time defining new variables and lose all the "time-saving" benefits provided by the computer.

Objective A: Size of a Vector and One-Dimensional Arrays

Consider an employer with 1,000 employees who wants to keep track of each employee's salary. Suppose the employer wants to sort the salary from lowest to highest. Let us see how we are going to organize the data so that we can easily perform the task of sorting the salary.

Employee Number	Employee's Name	Salary
1.	John	$ 36,000.00
2.	Mary	$ 73,500.00
3.	Debbie	$ 42,850.00
4.	Peter	$112,358.00
5.	Christine	$ 23,567.00
⋮	⋮	⋮
1,000.	Laura	$ 19,210.00

We could define 1,000 different variables for each employee's salary, such as {JohnsSalary, MarysSalary, DebbiesSalary,…}. Not only do we have to make sure that all 1,000 variables are distinct, we also want to be able to reference them with ease when working with them.

The solution to this problem is to define a new and different type of variable. Instead of naming each item in the list of data, we simply name the entire list. We now no longer need a thousand different names, we only need one. For our problem, let us name this list "**Salary**." We write this list as one long column of values that has a total of 1,000 rows surrounding them with brackets, i.e.,

Definition: Let n be a positive integer. An ordered collection of data,

$$v_1, v_2, \ldots, v_{n-1}, v_n$$

is called a **vector** of length (size) n. Each individual datum v_i is called a component or element of the vector. The subscript is used to signify which entry in the vector we are referring to.

$$\mathbf{Salary} = \begin{bmatrix} 36000.00 \\ 73500.00 \\ 42850.00 \\ 112358.99 \\ 23567.00 \\ \vdots \\ 19210.00 \end{bmatrix}$$

The vector is also called a column vector when written in column form:

$$\mathbf{v} = \begin{bmatrix} v_1 \\ v_2 \\ \vdots \\ v_{n-1} \\ v_n \end{bmatrix}$$

or a row vector, when written in row form:

$$\mathbf{v} = [v_1, v_2, \ldots, v_{n-1}, v_n]$$

In programming, we call a vector an array, more specifically **one-dimensional array**. The array is given a name and each individual element within the array is referenced using a subscript, also called an index.

Some comments about notations:

1. In mathematics, we generally use single bold letters, such as **v** or **u** to name vectors while in computer programming, we use variable names such as **Salary** or **Grade** to name arrays.

2. In mathematics, for an integer k, we use the notation v_k to refer to the kth element of vector **v**, in computer programming we use a subscript v(k) to refer to the kth element within the array. The subscript (or index) in an array is enclosed in parentheses after the array name.

Throughout this chapter we will use the terms vector and one-dimensional array interchangeably. We shall also use the notations v_k and v(k) interchangeably. Just keep in mind that in programming only array subscript or index is valid.

For example, if we want to refer to the fourth entry or Peter's salary in our list, we refer to it as $Salary_4$. We could have named the vector **v** and refer to Peter's salary as v_4. However, in programming we have to use the notation Salary(4) in the array.

Now, how do we store this list of salaries in the computer so that we can work with the list? We can store each of these 1,000 salaries in 1,000 consecutive storage locations. The entire block of 1,000 storage spaces is given a name, **Salary** in our example, and an individual element can be referred to using its position in the list (array). We'll discuss more about how to work with arrays in programming in the next section.

Note that vectors can contain numeric data or non-numeric data such as character strings.

Example 1: Given a row vector $\mathbf{v}=[3,7,-1,6]$, find v_2 or v(2).

  Solution: v(2) = 7 since the second entry or element of the row vector is 7.

Example 2: Given a column vector $\mathbf{v}=\begin{bmatrix}2\\0\\5\end{bmatrix}$, find v(3).

  Solution: v(3) = 5 since the third entry or element of the column vector is 5.

Size of a Vector

The **size** (length) of a vector or one-dimensional array is the number of elements in the vector. Vectors can come in any size desired (number of rows for a column-vector, or number of columns for a row-vector). Mathematically, there are no limitations when we work with vectors. However, when working with arrays in computers, there are limitations since only a fixed and limited amount of storage is available.

Example 3: *Give the size or length of each of the following row vectors:*

  a) **S** = [12, -10, 7, 3, 4, 1, 16]
  b) **T** = [0, 2, 4, 6,8]
  c) **U** = [0]:
  d) **V** = ["car", "house", "garage", "dog"]

Solution: a) **S** has 7 elements, so it is of length 7.
b) **T** has 5 entries, so we say it is of length 5.
c) **U** has only 1 entry, hence we say it is of length 1.
d) **V** has 4 entries, so we say that it is of length 4.

Example 4: Give the size or length of each of the following column vectors:

a) $\mathbf{v} = \begin{bmatrix} 4 \\ 2 \\ 9 \end{bmatrix}$ b) $\mathbf{w} = \begin{bmatrix} 2 \\ -1 \\ 4 \\ 6 \\ -8 \\ 3 \end{bmatrix}$

Solution: a) The vector **v** has 3 elements, so it is of length 3.
b) The vector **w** has 6 entries, so we say it is of length 6.

Objective B: Equality of Vectors

Two vectors (one-dimensional arrays) are said to be equal if the vectors are the same size and if all the corresponding elements in one vector are exactly the same as the corresponding elements in the other vector. For example: **v**=[1,0,1] and **w**=[1,0,1] are equal vectors because they both contain three entries and the three entries are 1, 0, and 1, respectively. So we write

$$\mathbf{v} = \mathbf{w}.$$

However, **v**=[1,0,1] and **m**=[1,0,1,0] are not equal vectors since the second vector **m** has an extra entry. Here we write

$$\mathbf{v} \neq \mathbf{m}.$$

Example 5: Determine which of the row-vectors **s**, **t**, **u**, **v**, and **w** defined below are equal.

s = [0, 1, -1, 3, 4, 21, 5.3]
t = [0, 1, -1, 3, 4, 21, 5.3, 6]
u = [5.3, 21, 4, 3, -1, 1, 0]

$\mathbf{v}$ = [0, 1, -1, 3, 4, 21, 5.3]
$\mathbf{w}$ = [0, 1, -1, 3, 4, 21]

Solution: First we note that the vectors **s**, **u** and **v** all have the same number of elements (7), while **t** and **w** have eight (8) and six (6) elements respectively. Thus, only **s**, **u** and **v** have a possibility of being equal to each other.

If we compare **s** and **u** we see that they both have the same elements, but not in the same locations. These two vectors are not equal to each other.

Finally, when we compare **s** with **v**, we see that they both have the same length as well as the same elements located in the same positions, therefore, these two vectors are equal and we write this mathematically as **s** = **v**.

As mentioned before, vectors or arrays are not limited to just containing numeric data. Arrays can also contain character data and the same definition of equality would apply.

Consider the following vectors:

$\mathbf{u}$ = ["Joe", "Mary", "Sal", "Liz", "Sam", "Donna"]
$\mathbf{v}$ = ["Joe", "Mary", "Sal", "Liz", "Sam", "Donna", "Ted"]
$\mathbf{w}$ = ["Joe", "Mary", "Sal", "Liz", "Sam", "Donna"]

We say the vector **u** equals vector **w** and that **v** is not equal to either vector **u** or vector **w** since it has a different number of elements. Also note that **u**, **v** and **w** are all character arrays.

EXERCISES 5.1

Give the size of each of the vectors.

1. $[-1 \quad 3 \quad 0 \quad -5]$

2. $[0 \quad -2 \quad 5 \quad 9 \quad -1 \quad 3]$

3. $\begin{bmatrix} 1 \\ -4 \\ -8 \\ 5 \end{bmatrix}$

4. $[1 \quad -4 \quad 9 \quad -5 \quad 0 \quad -3 \quad 8]$

5. $\begin{bmatrix} 1 \\ 2 \\ 3 \end{bmatrix}$

Determine if each of the following pairs of vectors are equal or not

6. $[3 \quad 5 \quad 2]$; $[3 \quad 5 \quad 2]$

7. $[3 \quad 7 \quad -1 \quad 9]$; $[3 \quad 9 \quad 7 \quad -1]$

8. $\begin{bmatrix} -3 \\ 5 \\ 0 \end{bmatrix}$; $\begin{bmatrix} 5 \\ 0 \\ -3 \end{bmatrix}$

9. $\begin{bmatrix} 0 \\ 1 \end{bmatrix}$; $\begin{bmatrix} 0 \\ 1 \end{bmatrix}$

10. $[1 \quad 0]$; $[0 \quad 1]$

Section 5.2: Vector Arithmetic

In addition to using vectors to store data in a convenient way, we would like to be able to perform mathematical computations on them. For example, let the vector **J** represent household expenses (we'll only include rent, food, utilities, and transportation) for the month of January and **F** represent the household expenses for the month of February.

$$\begin{array}{cccc} \text{rent} & \text{food} & \text{util} & \text{trans} \end{array} \qquad \begin{array}{cccc} \text{rent} & \text{food} & \text{util} & \text{trans} \end{array}$$
J = [900, 450, 150, 200] and **F** = [900, 500, 200, 180]   Each entry is in dollars.

Suppose at the end of February we want to find what the total expenses were for the two months in each category. We would add the corresponding entries in each vector.

We have to emphasize here that to be able to add the corresponding entries in the vectors **J** and **F**, the two vectors must have the same number of elements. In this case, **J** and **F** do have the same number of elements.

Focusing on numeric vectors, it makes sense that some of the rules of arithmetic related to real numbers should be applicable.

Objective A: Addition and Subtraction of Vectors

Given a set of vectors that are all the same size, we define the operations of addition and subtraction. If vectors are not the same size, then we cannot perform addition or subtraction on these vectors. In this section we will only show the row vectors, but everything we show here applies in the same way to column vectors.

To add or subtract two vectors, we add or subtract their corresponding elements.

Definition: Given two vectors $\mathbf{u} = [u_1, u_2, u_3, \ldots, u_{n-2}, u_{n-1}, u_n]$ and
$\mathbf{v} = [v_1, v_2, v_3, \ldots, v_{n-2}, v_{n-1}, v_n]$ with vectors **u** and **v** having the same number of elements, n, then we define addition or subtraction as follows:

$$\mathbf{u} + \mathbf{v} = [(u_1 + v_1), (u_2 + v_2), (u_3 + v_3), K, (u_{n-2} + v_{n-2}), (u_{n-1} + v_{n-1}), (u_n + v_n)]$$

and

$$\mathbf{u} - \mathbf{v} = [(u_1 - v_1), (u_2 - v_2), (u_3 - v_3), K, (u_{n-2} - v_{n-2}), (u_{n-1} - v_{n-1}), (u_n - v_n)]$$

Example 1: *Given the following two vectors,* **u** = [2, -1, 3] *and* **v** = [5, 2, -1], *find* **u** + **v**, **u** − **v**, *and* **v** − **u**.

Solution:
$$\mathbf{u} + \mathbf{v} = [(2 + 5), (-1 + 2), (3 + (-1))]$$
$$= [7, 1, 2]$$

$$\mathbf{u} - \mathbf{v} = [(2 - 5), (-1 - 2), (3 - (-1))]$$
$$= [-3, -3, 4]$$

$$\mathbf{v} - \mathbf{u} = [(5 - 2), (2 - (-1)), (-1 - 3)]$$
$$= [3, 3, -4]$$

Example 2: *Given the following two vectors,* **u** = $[2, 0]$ *and* **v** = $[0, -3]$, *find* **u** + **v**, **u** − **v**, *and* **v** − **u**.

Solution: $\mathbf{u} + \mathbf{v} = [2, 0] + [0, -3] = [2+0, 0+(-3)] = [2, -3]$

$$\mathbf{u} - \mathbf{v} = [2, 0] - [0, -3] = [2-0, 0-(-3)] = [2, 3]$$

$$\mathbf{v} - \mathbf{u} = [0, -3] - [2, 0] = [0-2, -3-0] = [-2, -3]$$

Example 3: *Given the following two vectors,* **u** = [2, 4, 8, 16] *and* **v** = [1, 4, 9, 16], *find* **u** + **v**, *and* **u** − **v**.

Solutions:
$$\mathbf{u} + \mathbf{v} = [(2+1), (4+4), (8+9), (16+16)]$$
$$= [3, 8, 17, 32]$$

$$\mathbf{u} - \mathbf{v} = [(2-1), (4-4), (8-9), (16-16)]$$
$$= [1, 0, -1, 0]$$

Now, for our household expense example above, to find the sub-total expenses for the months of January and February, we can add the two vectors as follows:

J = [900, 450, 150, 200] and **F** = [900, 500, 200, 180]

J + **F** = [(900+900), (450+500), (150+200), (200+180)]
    = [1800, 950, 350, 380]

To interpret the result of this addition, the total sub-expenses for January and February are: rent = $1,800; food = $950; utilities = $350; and transportation = $380.

If we want to subtract the two vectors, we have:

F - **J** = [(900−900), (500−450), (200−150), (180−200)]
    = [0, 50, 50, −20]

To interpret the result of the difference of the two vectors, if the result is positive, it means there is an increase; if the result is negative, it means there is a decrease; and if the result is 0, the monthly expense has not changed. So for our example, the rent stays the same, there is an increase of $50 in the food expense, an increase of $50 in utilities, and a decrease of $20 in transportation.

Objective B: Multiplication of Vectors

Multiplying vectors is a bit more complicated. We can multiply a vector by a real number or we can multiply two vectors together. The first is called scalar multiplication while the second is called inner product or dot product.

Scalar Multiplication:

Definition: **Scalar multiplication** of a vector is the product of a vector with any real number. That is, multiply each element in the vector **v** by the scalar *a*.

For example, given **v** = [1, 4, 9, 16, 25], then the scalar product 3**v** is as follows:

3**v**=3*[1, 4, 9, 16, 25]= [3*1, 3*4, 3*9, 3*16, 3*25]=[3, 12, 27, 48, 75]

and

$$-\mathbf{v} = -1*[1,\ 4,\ 9,\ 16,\ 25]$$
$$= [-1*1,\ -1*4,\ -1*9,\ -1*16,\ -1*25]$$
$$= [-1,\ -4,\ -9,\ -16,\ -25]$$

Example 1: Let $\mathbf{v} = [2,\ -3,\ 5]$ and $\mathbf{u} = [-1,\ 4,\ 2]$. Find $2\mathbf{v}$, $3\mathbf{u}$, and $2\mathbf{v} + 3\mathbf{u}$.

Solution: Note first that $2\mathbf{v}$ is the same as $2*\mathbf{v}$.

Definition: In general, the **scalar product** of a scalar (number) "a" with a vector

$$\mathbf{v} = [v_1, v_2, v_3, \ldots, v_{n-2}, v_{n-1}, v_n]\text{, is defined by:}$$

$$a\mathbf{v} = a[v_1, v_2, v_3, \ldots, v_{n-2}, v_{n-1}, v_n] = [av_1, av_2, av_3, \ldots, av_{n-2}, av_{n-1}, av_n]$$

$$2\mathbf{v} = [2(2),\ 2(-3),\ 2(5)] = [4,\ -6,\ 10]$$

$$3\mathbf{u} = [3(-1),\ 3(4),\ 3(2)] = [-3,\ 12,\ 6]$$

$$2\mathbf{v} + 3\mathbf{u} = [4,\ -6,\ 10] + [-3,\ 12,\ 6] = [1,\ 6,\ 16]$$

Example 2: Let $\mathbf{v} = [4,\ -3]$ and $\mathbf{u} = [-1,\ 2]$. Find $-2\mathbf{v}$, and $-\mathbf{u}$.

Solution: $-2\mathbf{v} = -2*[4,\ -3] = [-2(4),\ -2(-3)] = [-8,\ 6]$

$-\mathbf{u} = -1*[-1,\ 2] = [-1(-1),\ -1(2)] = [1,\ -2]$

Vector Products (Inner Product):

We now define a multiplication rule for any two vectors with the same number of elements. There are several types of vector multiplication. The most common is the **Inner**, also called the **Dot**, product of two vectors. The other forms of vector multiplication are beyond the scope of this course and will not be discussed further.

The inner product of two vectors is unique in that the resulting quantity is no longer a vector, but is now a scalar or regular number. The inner product is an operation that transforms a vector into a scalar.

Definition: The **inner product** or **dot product** between two vectors

$\mathbf{u} = [u_1, u_2, u_3, \ldots, u_{n-2}, u_{n-1}, u_n]$ and $\mathbf{v} = [v_1, v_2, v_3, \ldots, v_{n-2}, v_{n-1}, v_n]$ is

defined as: $\mathbf{u} \cdot \mathbf{v} = u_1 v_1 + u_2 v_2 + \ldots + u_{n-1} v_{n-1} + u_n v_n$,

the sum of all the products of the corresponding elements.

The inner product can be illustrated using a simple example. We define two vectors **u** and **v**. Let **u** be a vector containing all the hourly wages of the employees at a business and **v** contain how many hours each employee worked in a given week. Then the product, $\mathbf{u} \cdot \mathbf{v}$, gives each employee's hourly wage multiplied by their hours for the week. Add the products together for all the employees. This is the total payroll for the week. The inner product is an easy way to express and compute this.

Assume there are 7 employees in this company. Let vector **u** be the vector that contains the 7 employees' hourly wage and let vector **v** be the vector that contains the number of hours each employee worked during the week:

u = [5.00, 7.50, 8.00, 7.50, 9.00, 12.00, 8.25] and
v = [35, 32, 25, 40, 35, 20, 30]

The unit of measure in each entry of vector **u** is dollars per hour while each entry in vector **v** is hours. When we multiply the hourly rate with the number of hours, the net result is dollars.

To find the inner product or dot product of these two vectors, we multiply the corresponding elements in each vector and then add the products.

$\mathbf{u} \cdot \mathbf{v}$ = 5.00*35 + 7.50*32 + 8.00*25 + 7.50*40 + 9.00*35 + 12.00*20 + 8.25*30
= 175.00 + 240.00 + 200.00 + 300.00 + 315.00 + 240.00 + 247.50
= 1717.50 dollars

Example 3: Given vectors **u**=[2, 4, 6] and **v**=[−1, 3, −2], find the dot product of **u** and **v**.

Solution: $\mathbf{u} \cdot \mathbf{v} = (2)*(-1) + (4)*(3) + (6)*(-2) = -2 + 12 - 12 = -2$

We multiply each of the corresponding elements in each vector together and then add the resulting products to obtain the scalar value of –2.

Example 4: Given vectors **w** = [5, -3, 6] and **v** = [-3, 0, 4], find **w·v** and **w·w**.

Solution: $\mathbf{w}\cdot\mathbf{v} = (5)*(-3) + (-3)*(0) + (6)*(4) = -15 + 0 + 24 = 9$

$\mathbf{w}\cdot\mathbf{w} = (5)*(5) + (-3)*(-3) + (6)*(6) = 25 + 9 + 26 = 70$

Sometimes we compute the inner product between a row vector and column vector. The important thing to remember in such a case is that the number of columns in the row vector must equal the number of rows in the column vector in order for us to be able to perform the inner product.

To illustrate this, let **u**=[2, 4, 6] and $\mathbf{v} = \begin{bmatrix} -1 \\ 0 \\ 3 \end{bmatrix}$, notice that there are 3 columns in vector **u** and there are 3 rows in vector **v**. When we perform the inner product **u·v**, we get

$\mathbf{u}\cdot\mathbf{v} = (2)*(-1) + (4)*(0) + (6)*(3) = -2 + 0 + 18 = 16$

If the number of columns in the first vector is not the same as the number of rows in the second vector, then the inner product cannot be performed.

Examples 5: Given vectors **w** = [5,–3, 6] and $\mathbf{v} = \begin{bmatrix} -3 \\ 0 \\ 1 \end{bmatrix}$, find **w·v**.

Solution: $\mathbf{w}\cdot\mathbf{v} = (5)*(-3) + (-3)*(0) + (6)*(1) = -15 + 0 + 6 = -9$

Examples 6: Given vectors **w** = [2, 7, 9] and $\mathbf{v} = \begin{bmatrix} -6 \\ 2 \end{bmatrix}$, find **w·v**.

Solution: Since **w** has 3 columns and **v** has only 2 rows, we cannot perform inner product in this problem.

Objective C: One-Dimensional Arrays and Computers

By combining the convenience of our one-dimensional array with the computational power of the computer, we can process large amounts of data easily and quickly. To understand how we work with arrays on computers, let's consider the following example.

Suppose we want to find out the average temperature over 10 days. We could enter each of the temperatures and name each as temp1, temp2, temp3, temp4, and so on…until temp10. This is fine if we are dealing with a small number of temperatures. However, we can name all the 10 memory locations where the temperatures are stored using a one-dimensional array called **temp**. We can then use the subscripts of the array to refer to the individual storage locations by specifying its position in the array. So for our example

temp1 temp2 temp3 temp4 temp5 temp6 temp7 temp8 temp9 temp10

can be renamed using an array as follows:

temp(1) temp(2) temp(3) temp(4) temp(5) temp(6) …

Here **temp** is the name of the array and the integer inside the parentheses is called the index (subscript) of the array **temp**. Instead of the following code of reading in the 10 temperatures and storing them in each of the 10 memory locations,

READ temp1, temp2, temp3, temp4, temp5
READ temp6, temp7, temp8, temp9, temp10
         •
         •
         •
DATA   65, 45, 67,68,59
DATA   65, 63, 75,56, 70

the same effect can be achieved by using a loop and the array **temp**:

FOR K = 1 TO 10
    READ   TEMP(K)
NEXT K
         •
         •
         •

⋮

DATA   65, 45, 67, 68, 59
DATA   65, 63, 75, 56, 70

The array **temp** is a list of data of the same type (either numeric or string) referred to by the single variable name **temp.**

After the program segment is executed, the array **temp** can be pictured as the following:

65	45	67	68	59	65	63	75	56	70
temp(1)	temp(2)	temp(3)	temp(4)	temp(5)	temp(6)	temp(7)	temp(8)	temp(9)	temp(10)

Each of the 10 elements of the array is also called a ***subscripted variable***. Before a subscripted variable can be used in a program, the array must be declared. This means the computer is 'instructed' to set aside the number of storage locations to accommodate the number of data.

The **DIM** (short for dimension) statement is used to declare the number of storage spaces allocated to hold the elements of the array. The word dimension in this context means the number of locations to reserve or the number of elements in the array. Don't confuse it with the word dimension in one-dimensional or two-dimensional arrays (We will discuss this more in objective D in section 5.4).

In the example that we have on temperatures, the declaration can be done as follows:

> **DIM temp (1 TO 10)**

This statement tells the computer to allocate 10 consecutive memory spaces to hold the 10 elements of the array named **temp**, and the subscripts will be indicated in a parentheses, with temp(1) indicating the first temperature, temp(2) the second temperature, and so on.

If we have more than one array, then we need to declare each of the arrays:

DIM *variable*1$\left(L_1 \text{ TO } U_1\right)$**, *variable*2**$\left(L_2 \text{ TO } U_2\right)$**, ...**
> where ***variable*1, *variable*2, . . .** are the array names and
> $L_1, U_1, L_2, U_2,...$ are the lower and upper subscripts allowed for each array. The computer will allocate the number of memory locations needed for each array.

OR

DIM *variable*1(U_1), *variable*2(U_2), . . .

        where ***variable*1, *variable*2, . . .** are the array names and $U_1, U_2, ...$ are upper subscripts allowed for each array. The computer will allocate the number of memory locations needed for each array and assume that the lower subscripts are 0 for each array.

Example of loop:

```
REM Introduction to arrays
CLS
DIM num(1 TO 10)
FOR i = 1 TO 10
  INPUT "enter a number"; num(i)
NEXT i

FOR j = 1 TO 10
    PRINT num(j)
 NEXT j
 END
```

The above code asks the user to enter a number and store that number in the array called num. The array has been declared to store up to 10 numbers. After all 10 numbers are entered and stored, then the program tells the computer to print the 10 numbers. Since the PRINT statement does not have a semi-colon or comma, each of the 10 numbers will be printed on a new line.

Some notes on arrays:

- If a subscripted array is used in a program before the array is declared, then the array will automatically be dimensioned with lower bound of 0 and upper bound of 10. However, it is good programming practice to always declare the arrays.

- When an array element is used in a program, the subscript could be a numeric constant, variable, or expression. When an expression is used, the value will be rounded to an integer and checked against the array's allowable subscript values. If it is not within the allowable values, then an error message "subscript out of range" will be displayed.
- When declaring an array, if there is only a numeric constant in the parentheses, then the range of subscript will be from 0 to that numeric constant. For example, **DIM temp(5)**,

means there are 6 memory locations allocated for this array, the smallest one is 0 and the largest one is 5.

Objective D: Geometric Interpretation of Vectors (Optional)

Vectors can also be looked at from a geometric perspective. Many of the definitions for vector addition, subtraction and multiplication can be understood better if we look at how vectors are developed geometrically.

Vector Addition:

Consider a vector **v** with 2 elements i.e. **v** = [2, 3]. We interpret the first elements of the vector "2" as the "x" coordinate and the second element "3" as the "y" coordinate, respectively, of a point in a Cartesian coordinate system. See the figure below. Now draw a directed line segment from the origin to the point (2, 3). A directed line segment is an arrow with a tail that begins at the origin and an arrowhead that terminates at the point (2, 3). Vectors can be interpreted as the net direction we have moved away from the origin. In this example, we moved 2 units in the x-direction and 3 units in the y-direction.

Now consider another vector **u** = [2, 1] and also represent this vector on our graph. If we add the vectors together, as we've shown above, we would obtain a new vector **w** = [2+2, 1+3] = [4,4].

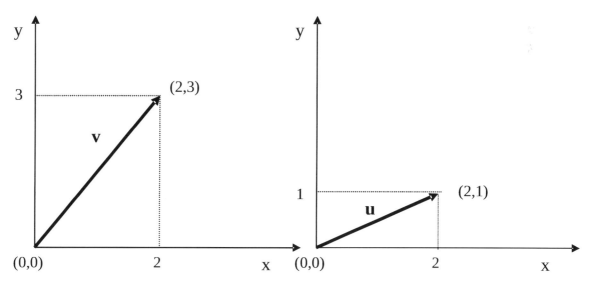

Now notice what happens if we move the tail of vector **u** to the head of vector **v**. This would be the distance we would have walked after we were at the head of vector **v**. Notice how the head of this vector terminates at exactly the same location as our vector **w**. Vector addition is nothing more than putting the tail of one vector at the head of the other and then drawing a vector from the origin to the head of the second vector.

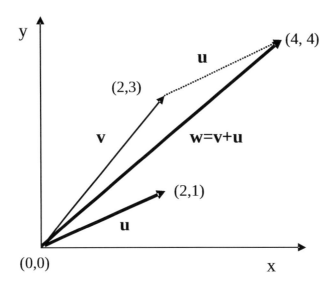

Vector Subtraction:

Subtraction is performed in a similar way. Given the same two vectors we defined above, we find the difference **v-u**. Recall that subtraction is equivalent to adding the negative of the number we are subtracting. If we apply the same rule to our vectors, then **v-u** is equivalent to $\mathbf{v} + (-\mathbf{u}) = [2, 3] + [-2, -1] = [0, 2]$.

Drawing the vector **–u** on our graph and then performing the same operation we did geometrically for the addition of two vectors. Next, place the tail of vector **–u** at the head of **v** to obtain the vector **v + (–u)**. Notice that this is the same vector we obtain by performing the subtraction of vectors as was done earlier in this chapter.

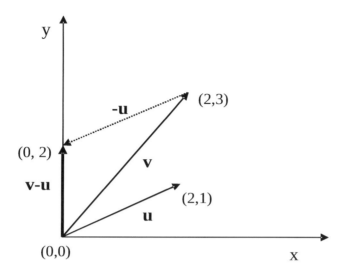

Vector Multiplication by a Scalar:

We can also consider scalar multiplication of a vector in a geometric way. Multiplying by a scalar quantity *a* either stretches the vector by a factor of *a*, if *a* is greater than one ($a > 1$), or it contracts the vector by a factor of *a*, if *a* is between one and zero. $0 < a < 1$. If *a* is negative, we are simply stretching or contracting the vector by a value equivalent to ABS(*a*), the absolute value of *a*, and then flipping the vector in the opposite direction

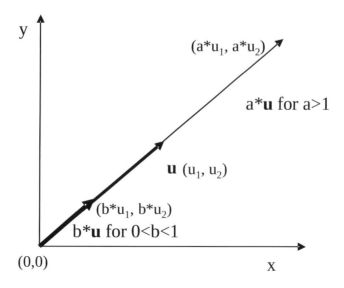

Vector Multiplication – Inner Product:

The inner product can best be understood by looking at the product of vector **u** with itself. Thus, we are multiplying **u·u** . If we draw our vector **u** and then drop a line perpendicular from the tip of the vector to the x-axis, we see that we have drawn a right triangle as illustrated in the figure below. Notice that the square of the length of the hypotenuse of this triangle is equal to our inner product. Recall from above that for a vector of size 2, . Thus $\mathbf{u} \cdot \mathbf{u} = u_1^2 + u_2^2$ the inner product of a vector with itself represents the length of that vector squared. Multiplying two different vectors together using the inner product is a bit more complicated to understand geometrically and is beyond the scope of this text.

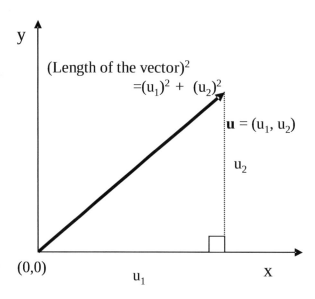

If we move beyond a vector of "dimension" 2, all the analogies follow, but in a more abstract (difficult or impossible to visualize) n-dimensional space. Here we use the word dimension to mean what we do in an ordinary sense of the word. We should note that there are some similarities between the size of a vector/array and the "dimension" of such a vector, but we will not go into details, as that may just confuse the reader.

EXERCISES 5.2

Add or subtract the following two vectors.

1. $[2, 5] + [3, 1]$
2. $[-1, 3, 4] + [2, 7, -5]$
3. $[3, -2] - [4, -1]$
4. $[1, -3, 8] - [-3, 6, -2]$

5. $[6, -3] - [-7, 4]$

Let $a = 2$ and $b = 3$ be scalars and vectors $\mathbf{u} = [4, -2, 3]$, $\mathbf{v} = [-1, 7, 5]$, and $\mathbf{w} = [-3, 10, -7]$. Perform the indicated operations.

6. $a\mathbf{u}$
7. $b\mathbf{v}$
8. $a\mathbf{u} + b\mathbf{v}$
9. $\mathbf{v} - \mathbf{u}$
10. $b\mathbf{v} - \mathbf{u}$
11. $\mathbf{v} + \mathbf{w}$
12. $\mathbf{u} \cdot \mathbf{v}$
13. $2\mathbf{w} \cdot \mathbf{v}$
14. $\mathbf{u} \cdot \mathbf{w}$
15. $a(\mathbf{u} + \mathbf{w})$

Find the inner product of the two vectors. If it is not possible, state why.

16. $[2, -4] \cdot \begin{bmatrix} -1 \\ 3 \end{bmatrix}$

17. $[-4, 2, 7] \cdot \begin{bmatrix} -3 \\ 5 \end{bmatrix}$

18. $[-1, 7, -3] \cdot \begin{bmatrix} -2 \\ 8 \\ 1 \end{bmatrix}$

19. $[-5, 9] \cdot \begin{bmatrix} 1 \\ -6 \\ 3 \end{bmatrix}$

20. $[3, -2, 7, -9] \cdot \begin{bmatrix} 1 \\ 4 \\ 0 \\ -1 \end{bmatrix}$

Section 5.3: Two-Dimensional Arrays: Matrices

One-dimensional arrays (vectors) provide us with a convenient way to organize large single lists of data. However, we frequently encounter lists that require more than one associated list. For example, in addition to keeping track of employees' salary, we also want to keep track of the amount of their last raise, their sick days, and the taxes withheld from them. We can identify each employee by a social security number or employee number. We could construct a new one-dimensional array for each of these quantities, but that would be quite cumbersome. Instead we could define a new type of list, called a **two-dimensional array** or **matrix** to handle this data type.

Objective A: Order of a Matrix and Two-Dimensional Arrays

A matrix is defined to be a single object that can store several lists together under a single name. It accomplishes this by simply appending vectors together thereby increasing the number of columns or rows.

A two-dimensional array (matrix) is an object that will have "m" rows and "n" columns.

Definition: Let m and n be positive integers. A two-dimensional array of data, **A** is defined as:

$$A = \begin{bmatrix} a_{11} & a_{12} & \cdots & a_{1n} \\ a_{21} & \ddots & & a_{2n} \\ \vdots & & & \vdots \\ a_{m1} & a_{m2} & & a_{mn} \end{bmatrix}$$

where each entry a_{ij} is placed in row i and column j. **A** is called a **matrix** or a **two-dimensional array**, i and j are referred to as the subscripts of the array. On a computer we reference an element of the array **A** as a(i,j). The plural of matrix is matrices.

For example, consider the list below

Employee number	Salary	Raise	Sick days
101	25000	1000	12
102	37500	1500	9
103	55000	2500	7
104	24750	1000	15
105	45000	3000	12

The above table is just a portion of a bigger list of employee numbers with their corresponding salaries, raises, and sick days. We could create vectors or one-dimensional arrays for each category: salary, raise, and sick days, but instead we organize all the data into a matrix where each column represents one of the categories. The first column contains the employee number, the second column the salary, the third column the raise received, and the fourth column the number of sick days. The rows represent the information for each employee.

Now if we put all this information into a matrix, we'll have the following:

$$\begin{bmatrix} 101 & 25000 & 1000 & 12 \\ 102 & 37500 & 1500 & 9 \\ 103 & 55000 & 2500 & 7 \\ 104 & 24750 & 1000 & 15 \\ 105 & 45000 & 3000 & 12 \end{bmatrix}$$

Just like we did with vectors, we need to give this matrix a name. We usually name a matrix with a capital letter of the alphabet. We can name this matrix as **A**. Hence,

$$\mathbf{A} = \begin{bmatrix} 101 & 25000 & 1000 & 12 \\ 102 & 37500 & 1500 & 9 \\ 103 & 55000 & 2500 & 7 \\ 104 & 24750 & 1000 & 15 \\ 105 & 45000 & 3000 & 12 \end{bmatrix}$$

Matrices are typically represented by a bold capital letter from the beginning of the alphabet, while vectors were given bold lower-case letters near the end of the alphabet. Although this is not always true, it is a convention that is frequently followed to help distinguish more easily between the two mathematical objects.

If a matrix has m rows and n columns, then we say that the matrix is of **order** $m \times n$ (read m by n).

The order of matrix **A** above is 5×4 since there are 5 rows and 4 columns.

Some clarifications regarding the word dimension are in order here. Up until now, we talked about a one-dimensional array which is essentially a row or a column vector. The word dimension in one-dimensional array means the array has one subscript and in two-dimensional array means there are two subscripts. Two-dimensional arrays will be can be thought of as being tables that have rows as well as columns. So a two-dimensional array of order 5×3 means it has five rows and three columns. And the subscripts look like a(m, n) where a is the array name, m refers to the row number while n refers to the column number, and a(m, n) refers to the element in the mth row and n^{th} column.

Example 1: *What is the order of each of the following matrices?*

$$a) \mathbf{B} = \begin{bmatrix} 1 & 2 \\ 3 & 4 \\ 5 & 6 \end{bmatrix} \quad b) \mathbf{C} = \begin{bmatrix} 4 & 7 & 9 & 3 \\ -5 & 0 & 1 & 3 \end{bmatrix}$$

Solution: a) The order of matrix **B** is 3×2.
b) The order of matrix **C** is 2×4.

Let us look at the example matrix **A** again. Identify a_{13} or $a(1,3)$. The subscript 13 refers to the entry in the first row and third column. That means it is 1000, which represents the raise for employee number 101.

Example 2: *Identify the elements b_{22} or $b(2,2)$ and c_{14} or $c(1,4)$ in the matrices given.*

$$\mathbf{B} = \begin{bmatrix} 1 & 2 \\ 3 & 4 \\ 5 & 6 \end{bmatrix} \quad \mathbf{C} = \begin{bmatrix} 4 & 7 & 9 & 3 \\ -5 & 0 & 1 & 3 \end{bmatrix}$$

Solution: b_{22} or $b(2,2) = 4$ and c_{14} or $c(1,4) = 3$

Objective B: Equality of Matrices

Just as we wanted to compare two one-dimensional arrays, we also want to compare two two-dimensional arrays. To do this, we need to define when two matrices are equal.

Definition: Given two matrices of the same order,

$$A = \begin{bmatrix} a_{11} & a_{12} & \cdots & a_{1n} \\ a_{21} & \ddots & & a_{2n} \\ \vdots & & & \vdots \\ a_{m1} & a_{m2} & & a_{mn} \end{bmatrix} \text{ and } B = \begin{bmatrix} b_{11} & b_{12} & \cdots & b_{1n} \\ b_{21} & \ddots & & b_{2n} \\ \vdots & & & \vdots \\ b_{m1} & b_{m2} & & b_{mn} \end{bmatrix}, \text{ the two matrices are said to be }$$

equal if $a_{ij} = b_{ij}$ for all $1 \leq i \leq m$ and $1 \leq j \leq n$.

Two matrices are said to be equal if they both have the same order (or size) and all their corresponding elements are equal.

Example 3: Given the following matrices, find which matrices are equal to each other.

$$A = \begin{bmatrix} 1 & 2 \\ 4 & -3 \end{bmatrix}, B = \begin{bmatrix} 1 & 2 \\ 4 & -3 \end{bmatrix}, C = \begin{bmatrix} 1 & 4 \\ 2 & -3 \end{bmatrix}, D = \begin{bmatrix} 1 & 2 \\ 4 & -3 \\ 0 & 1 \end{bmatrix}, E = \begin{bmatrix} 1 & -7 & 3 \\ 6 & 0 & 2 \end{bmatrix}$$

Solution: Only matrices A, B and C are the same size (2×2), while D is 3×2 and E is 2×3. Thus, it is only possible for either A, B or C to be equal to one another. We notice that while they all have the same elements, 1, 2, 4, -3, only A and B have those elements in the same exact location. Thus, A and B are equal and C is not, i.e. A=B.

Example 4: Find the values of variables *x* and *y* that make the following equality true.

$$\begin{bmatrix} 3 & x \\ 9 & 7 \end{bmatrix} = \begin{bmatrix} 3 & 5 \\ y+1 & 7 \end{bmatrix}$$

Solution: Both matrices have the same order of 2×2, so we can compare the corresponding entries. If the matrices are equal, then corresponding entries must be equal to each other. Thus equating corresponding entries $3 = 3$, $x = 5$, $9 = y + 1$, $7 = 7$. The first and last equality are true. The second and third will be true if $x = 5$ and $y = 9 - 1$ or 8.

Note: In order for two matrices to be equal, they must both have the same order and their corresponding entries must be equal.

EXERCISES 5.3

What is the order of each of the following matrices?

1. $\begin{bmatrix} 3 & 0 & 9 \\ 1 & -3 & 2 \end{bmatrix}$

2. $\begin{bmatrix} 2 & -9 & 4 \\ 0 & 2 & 5 \\ 5 & 1 & -2 \end{bmatrix}$

3. $\begin{bmatrix} 8 & 3 & 1 & -2 \end{bmatrix}$

4. $\begin{bmatrix} 0 & 3 \\ 7 & 1 \\ -8 & 4 \\ 0 & -6 \\ 7 & -1 \end{bmatrix}$

Given the matrix $\mathbf{P} = \begin{bmatrix} 5 & 7 & 0 & 3 \\ 4 & 6 & 1 & 8 \\ 6 & 9 & 0 & 3 \\ 4 & 6 & 3 & 8 \end{bmatrix}$, *identify each of the following elements:*

5. $\mathbf{P}_{22}$

6. $\mathbf{P}_{43}$

7. $\mathbf{P}(4,1)$

8. $\mathbf{P}(1,3)$

Find the values of variables that make the following equality true. If the equality is impossible, state the reason.

9. $\begin{bmatrix} 6 & x-2 \\ 2y & 2 \end{bmatrix} = \begin{bmatrix} 6 & 5 \\ 8 & 2 \end{bmatrix}$

10. $\begin{bmatrix} 3x-1 & 2 & 4 \\ z+3 & 8 & w \end{bmatrix} = \begin{bmatrix} 5 & 2 & y+1 \\ 0 & 8 & 1 \end{bmatrix}$

11. $\begin{bmatrix} x & 1 & z \\ 2 & y & 3 \end{bmatrix} = \begin{bmatrix} 7 & 2 \\ 1 & 7 \\ 3 & 3 \end{bmatrix}$

12. $\begin{bmatrix} 3 & x \\ 4-y & 5 \end{bmatrix} = \begin{bmatrix} 3 & 2 \\ 1 & z \end{bmatrix}$

13. $\begin{bmatrix} x-2 & 8 & y-1 \\ 4 & 2z & 4 \end{bmatrix} = \begin{bmatrix} 2 & 2x & 2 \\ 4 & 10 & z-1 \end{bmatrix}$

Section 5.4: Matrix Arithmetic

Objective A: Addition and Subtraction of Matrices

Just like vectors, we can add or subtract two matrices. However, we must make sure that the matrices have the same order, i.e., they are both $m \times n$. Then we can add or subtract the matrices by adding or subtracting the corresponding elements from each matrix.

Examples: Given the matrices below, find $A + B$, $A - B$, and $A + C$.

$$A = \begin{bmatrix} 1 & 2 \\ 4 & -3 \end{bmatrix} \quad B = \begin{bmatrix} -1 & 4 \\ 6 & -5 \end{bmatrix} \quad C = \begin{bmatrix} 1 & 0 & 1 \\ 2 & 6 & -5 \end{bmatrix}$$

Solution:
$$A + B = \begin{bmatrix} 1 & 2 \\ 4 & -3 \end{bmatrix} + \begin{bmatrix} -1 & 4 \\ 6 & -5 \end{bmatrix} = \begin{bmatrix} 1+(-1) & 2+4 \\ 4+6 & -3+(-5) \end{bmatrix} = \begin{bmatrix} 0 & 6 \\ 10 & -8 \end{bmatrix}$$

$$A - B = \begin{bmatrix} 1 & 2 \\ 4 & -3 \end{bmatrix} - \begin{bmatrix} -1 & 4 \\ 6 & -5 \end{bmatrix} = \begin{bmatrix} 1-(-1) & 2-4 \\ 4-6 & -3-(-5) \end{bmatrix} = \begin{bmatrix} 2 & -2 \\ -2 & 2 \end{bmatrix}$$

$A + C$ cannot be done because they do not have the same order or size. A is 2×2 and C is 2×3

Objective B: Multiplication of Matrices

Because matrices have a structure that is different than real numbers, we'd expect some differences in the rules of arithmetic. Multiplying matrices is where this arises. Just as we had with vectors, there are different types of multiplication. The first is multiplying a matrix by a scalar. The second is when we multiply two matrices together.

Definition: **Scalar Multiplication**

Assume A is an m x n matrix and k is a real number. The scalar multiple of matrix A by a real number k, denoted by kA, is the matrix B where each of the entries of B are obtained by multiplying each of the entries of A by the real number k.

To illustrate using a 2×3 matrix, If $\mathbf{A} = \begin{bmatrix} a_{11} & a_{12} & a_{13} \\ a_{21} & a_{22} & a_{23} \end{bmatrix}$, then

$$\mathbf{B} = k\mathbf{A} = k\begin{bmatrix} a_{11} & a_{12} & a_{13} \\ a_{21} & a_{22} & a_{23} \end{bmatrix} = \begin{bmatrix} ka_{11} & ka_{12} & ka_{13} \\ ka_{21} & ka_{22} & ka_{23} \end{bmatrix}$$

Example: Given $\mathbf{A} = \begin{bmatrix} 2 & 7 & -3 \\ 5 & 9 & 6 \\ -1 & 4 & 0 \end{bmatrix}$ and $k = 6$, find $k\mathbf{A}$.

Solution: $k\mathbf{A} = 6\mathbf{A} = 6\begin{bmatrix} 2 & 7 & -3 \\ 5 & 9 & 6 \\ -1 & 4 & 0 \end{bmatrix} = \begin{bmatrix} 12 & 42 & -18 \\ 30 & 54 & 36 \\ -6 & 24 & 0 \end{bmatrix}$

Matrix Multiplication

To multiply two matrices together, we first must establish a consistent procedure for doing so. We shall define matrix multiplication as follows.

Definition: Given two matrices, **A** and **B**, we define **matrix multiplication** as the inner product or dot product of the row vectors of the first matrix with the column vectors of the second matrix and write the result in the (row,column) location in our new matrix. This definition of matrix multiplication automatically places some restrictions on the matrices we are multiplying. The number of elements in the rows of the first matrix, must equal the number of elements in the columns of the second matrix, otherwise the multiplication discussed above is not defined. This means that the number of columns in the first matrix must equal the number of rows in the second matrix.

Thus, suppose that **A** and **B** are two matrices and that **A** is an m × n matrix (m rows and n columns) and that **B** is a p × q matrix. In order for us to be able to multiply **A** and **B** together, **A** must have the same number of columns as **B** has rows (i.e., n = p). The product will be a matrix with m rows and q columns. To find the entry in row r and column c of the new matrix we take the "dot product" of row r of matrix **A** and column c of matrix **B** (pair up the elements of row r with column c, multiply these pairs together individually, and then add their products).

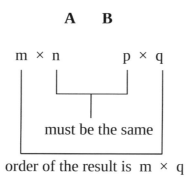

If n is not the same as p, then the product **AB** cannot be computed.

To illustrate this multiplication, suppose $\mathbf{A} = \begin{bmatrix} a_{11} & a_{12} \\ a_{21} & a_{22} \end{bmatrix}$ and $\mathbf{B} = \begin{bmatrix} b_{11} & b_{12} \\ b_{21} & b_{22} \end{bmatrix}$,

since both matrices have the same order 2×2, we can multiply the two matrices.

$$\mathbf{AB} = \begin{bmatrix} a_{11} & a_{12} \\ a_{21} & a_{22} \end{bmatrix} \begin{bmatrix} b_{11} & b_{12} \\ b_{21} & b_{22} \end{bmatrix}$$

The steps that we use to obtain the entries in the resulting matrix are:

1. We multiply the entries in row 1 of matrix **A** with the corresponding entries of column 1 of matrix **B** to obtain the row 1 column 1 entry of the result.

    This equals $a_{11}b_{11} + a_{12}b_{21}$.

2. We multiply the entries in row 1 of matrix **A** with the corresponding entries of column 2 of matrix **B** to obtain the row 1 column 2 entry of the result.

    This is $a_{11}b_{12} + a_{12}b_{22}$.

3. We multiply the entries in row 2 of matrix **A** with the corresponding entries of column 1 of matrix **B** to obtain the row 2 column 1 entry of the result.

   This is $a_{21}b_{11} + a_{22}b_{21}$.

4. We multiply the entries in row 2 of matrix **A** with the corresponding entries of column 2 of matrix **B** to obtain the row 2 column 2 entry of the result.

   This is $a_{21}b_{12} + a_{22}b_{22}$.

$$\mathbf{AB} = \begin{bmatrix} a_{11} & a_{12} \\ a_{21} & a_{22} \end{bmatrix} \begin{bmatrix} b_{11} & b_{12} \\ b_{21} & b_{22} \end{bmatrix} = \begin{bmatrix} a_{11}b_{11} + a_{12}b_{21} & a_{11}b_{12} + a_{12}b_{22} \\ a_{21}b_{11} + a_{22}b_{21} & a_{21}b_{12} + a_{22}b_{22} \end{bmatrix}$$

A specific illustration of matrix multiplication follows:

Suppose we have the matrices

$$A = \begin{bmatrix} 1 & 2 & 3 \\ 4 & 5 & 6 \end{bmatrix} \quad B = \begin{bmatrix} 1 & 2 \\ 2 & 5 \\ 4 & 9 \end{bmatrix}$$

Note that **A** is order 2 ×3 and **B** is order 3 × 2. Since the number of columns in **A** equals the number of rows in **B** (the second number in the order of the first matrix equals the first number in the order of the second matrix), we are able to find the matrix product **AB**. This new matrix will have order 2 × 2.

This procedure is also equivalent to treating the first row of **A** as a row-vector and the first column of **B** as a column vector and then computing the inner product of these two vectors and putting the result in the row one, column one position of **AB**.

$$\begin{bmatrix} 1 & 2 & 3 \end{bmatrix} \bullet \begin{bmatrix} 1 \\ 2 \\ 4 \end{bmatrix} = (1*1 + 2*2 + 3*4) = 17, \quad \mathbf{AB} = \begin{bmatrix} 17 & \# \\ \# & \# \end{bmatrix}$$

Then we'd take the first row of **A** and take its inner product with the second column of **B** and put the result in the row one column two position of **AB**.

$$[1\ 2\ 3] \bullet \begin{bmatrix} 2 \\ 5 \\ 9 \end{bmatrix} = (1*2+2*5+3*9) = 39, \qquad AB = \begin{bmatrix} \# & 39 \\ \# & \# \end{bmatrix}$$

Then move to the second row of **A** and perform the same process on the first and second column of **B**.

$$[4\ 5\ 6] \bullet \begin{bmatrix} 1 \\ 2 \\ 4 \end{bmatrix} = (4*1+5*2+6*4) = 38, \qquad AB = \begin{bmatrix} \# & \# \\ 38 & \# \end{bmatrix}$$

$$[4\ 5\ 6] \bullet \begin{bmatrix} 2 \\ 5 \\ 9 \end{bmatrix} = (4*2+5*5+6*9) = 87, \qquad AB = \begin{bmatrix} \# & \# \\ \# & 87 \end{bmatrix}$$

Putting everything together, we have:

$$AB = \begin{bmatrix} (1*1+2*2+3*4) & (1*2+2*5+3*9) \\ (4*1+5*2+6*4) & (4*2+5*5+6*9) \end{bmatrix} = \begin{bmatrix} 17 & 39 \\ 38 & 87 \end{bmatrix}$$

Example: *Given the following matrices*

$$A = \begin{bmatrix} 1 & 2 \\ 4 & -3 \end{bmatrix} \quad B = \begin{bmatrix} -1 & 4 \\ 6 & -5 \end{bmatrix} \quad C = \begin{bmatrix} 1 & 0 & 1 \\ 2 & 6 & -5 \end{bmatrix}$$

Perform matrix multiplication **BA** *and* **CA**. *If the multiplication cannot be performed, state the reason why.*

Solution: $BA = \begin{bmatrix} -1 & 4 \\ 6 & -5 \end{bmatrix}\begin{bmatrix} 1 & 2 \\ 4 & -3 \end{bmatrix} = \begin{bmatrix} -1(1)+4(4) & -1(2)+4(-3) \\ 6(1)+(-5)(4) & 6(2)+(-5)(-3) \end{bmatrix} = \begin{bmatrix} 15 & -14 \\ -14 & 27 \end{bmatrix}$

CA is not possible to perform since **C** is 2×3 and **A** is 2×2. The number of columns in matrix **C** is not the same as the number of rows in matrix **A**.

Associativity and Commutivity of Matrix Multiplication

Consider the two matrices:

$$A = \begin{bmatrix} 1 & 0 \\ 3 & 2 \\ 4 & -1 \end{bmatrix} \quad \text{and} \quad B = \begin{bmatrix} 0 & 2 & -1 \\ 3 & 1 & 0 \end{bmatrix}$$

A is a 3 x 2 matrix and B is a 2 x 3 matrix. Compute the matrix product AB which is a 3 x 3 and obtain:

$$AB = \begin{bmatrix} 0 & 2 & -1 \\ 6 & 8 & -3 \\ -3 & 7 & -4 \end{bmatrix}$$

If we now compute the matrix product BA which is a 2 x 2 matrix, we obtain:

$$BA = \begin{bmatrix} 2 & 5 \\ 6 & 2 \end{bmatrix}$$

Thus, we have shown that, in general, the matrix product **AB** ≠ **BA**. That is, matrix multiplication does not commute. Not only are the elements of the matrix different in this example, but the order of the resulting matrices is different as well.

Matrix multiplication, however, does obey the associative law. If we have three matrices **A**, **B** and **C** then:

$$A(BC)=(AB)C$$

The reader should verify this by doing the matrix multiplication on the above equation with A and B from the example above and use

$$C = \begin{bmatrix} 1 & 0 \\ 3 & -1 \\ 0 & -2 \end{bmatrix}$$

Objective C: Identity Matrix

Definition: An **identity matrix** is a square matrix of order $n \times n$ with entries of 1 on the main diagonal (upper left to lower right) and 0 for the remaining entries.

The **identity matrix, I,** has the following properties: For any matrix **A**, **AI = IA = A**.

Example: If $\mathbf{A} = \begin{bmatrix} a_{11} & a_{12} & a_{13} \\ a_{21} & a_{22} & a_{23} \\ a_{31} & a_{32} & a_{33} \end{bmatrix}$, then the identity matrix will be $\mathbf{I} = \begin{bmatrix} 1 & 0 & 0 \\ 0 & 1 & 0 \\ 0 & 0 & 1 \end{bmatrix}$ so that

$$\underset{\mathbf{A}}{\begin{bmatrix} a_{11} & a_{12} & a_{13} \\ a_{21} & a_{22} & a_{23} \\ a_{31} & a_{32} & a_{33} \end{bmatrix}} * \underset{\mathbf{I}}{\begin{bmatrix} 1 & 0 & 0 \\ 0 & 1 & 0 \\ 0 & 0 & 1 \end{bmatrix}} = \underset{\mathbf{I}}{\begin{bmatrix} 1 & 0 & 0 \\ 0 & 1 & 0 \\ 0 & 0 & 1 \end{bmatrix}} * \underset{\mathbf{A}}{\begin{bmatrix} a_{11} & a_{12} & a_{13} \\ a_{21} & a_{22} & a_{23} \\ a_{31} & a_{32} & a_{33} \end{bmatrix}} = \underset{\mathbf{A}}{\begin{bmatrix} a_{11} & a_{12} & a_{13} \\ a_{21} & a_{22} & a_{23} \\ a_{31} & a_{32} & a_{33} \end{bmatrix}}$$

Example 1: Find the 2×2 identity matrix, then for matrix $\mathbf{C} = \begin{bmatrix} 3 & 7 \\ -2 & 4 \end{bmatrix}$, show that

$$\mathbf{IC = CI = C}$$

Solution: Matrix **C** is a 2×2 matrix, therefore we are looking for a 2×2 matrix with 1's along the diagonal and 0's everywhere else, or

$$I = \begin{bmatrix} 1 & 0 \\ 0 & 1 \end{bmatrix}$$

We now verify this using matrix multiplication, i.e.,

$$\mathbf{IC} = \begin{bmatrix} 3 & 7 \\ -2 & 4 \end{bmatrix}\begin{bmatrix} 1 & 0 \\ 0 & 1 \end{bmatrix} = \begin{bmatrix} 3+0 & 0+7 \\ -2+0 & 0+4 \end{bmatrix} = \begin{bmatrix} 3 & 7 \\ -2 & 4 \end{bmatrix}$$

and

$$\mathbf{CI} = \begin{bmatrix} 1 & 0 \\ 0 & 1 \end{bmatrix}\begin{bmatrix} 3 & 7 \\ -2 & 4 \end{bmatrix} = \begin{bmatrix} 3+0 & 7+0 \\ 0-2 & 0+4 \end{bmatrix} = \begin{bmatrix} 3 & 7 \\ -2 & 4 \end{bmatrix}$$

We have shown that **IC = CI = C**

Example 2: Find the 3×3 identity matrix, then for matrix $\mathbf{D} = \begin{bmatrix} -2 & 1 & 4 \\ 4 & 8 & 15 \\ 3 & 0 & -5 \end{bmatrix}$, show that

ID = DI = D

Solution: Matrix **D** is a 3 × 3 matrix, therefore we looking for a 3 ×3 matrix with 1's along the diagonal and 0's everywhere else, or

$$\mathbf{I} = \begin{bmatrix} 1 & 0 & 0 \\ 0 & 1 & 0 \\ 0 & 0 & 1 \end{bmatrix}$$

As an exercise the reader should verify that **DI = ID = D.**

Objective D: Two-Dimensional Arrays and Computers

As we discussed above, a **two–dimensional array** is a collection of subscripted elements or "variables," each of which has two subscripts (separated by a comma). This two-dimensional array can be used to store the content in a table with a specified number of rows and columns.

To illustrate how we work with two-dimensional arrays on a computer, let's consider the following example. Suppose the following table represents the grades of each of the three tests for five students. We can make the rows represent the five students and the columns represent the three grades of each of these five students.

	test 1	test 2	test 3
Student 1	75	68	82
Student 2	90	75	88
Student 3	65	80	67
Student 4	99	78	90
Student 5	66	87	100

Suppose we declare this two-dimensional array as **test(5,3)**, the number 5 in the parentheses represents the number of rows in the table, while the number 3 represents the number of columns in the table. Just like the one-dimensional array, we must also declare two-dimensional arrays. One way of declaring a two-dimensional array is **DIM test(5,3)**. Another way of declaring this array is **DIM test(1 TO 5, 1 TO 3)**. Now if we want to refer to the second grade of the third student, then we simply want to know what is in the third row, second column of the table, which translates to the subscripted variable **test** (3,2). This grade is 80.

Some notes on two-dimensional arrays:
- A two-dimensional array variable can hold either numeric data or string data, but not both.
- A two-dimensional array uses nested loops. That is, a loop within a loop.
- We can extend the arrays to multi-dimensional arrays, as long as we are consistent with the data type.

What will the output of this program look like?
```
CLS
DIM TEST (1 TO 5, 1 TO 3)
FOR I = 1 TO 5
   FOR J = 1 TO 3
       READ TEST (I, J)
       PRINT TEST (I, J);
   NEXT  J
PRINT
NEXT I

DATA 90, 80, 70, 85, 75, 65, 72, 44, 55,55, 66, 77
DATA 99, 94, 99
```

Let us extend the program that we started above and do it for 10 students. Read the data that consists of the names of the 10 students as well as the test grades for each of the three tests that each student received. Then calculate the average of the each student. Place the result in an array called **ave**.

```
CLS
'Enter the students' names and grades
DIM names$(1 TO 10)
DIM grades(10, 3)
DIM ave(1 TO 10), total(1 TO 10)
FOR i = 1 TO 10
     READ names$(i)
```

```
      PRINT names$(i);
      total(i) = 0
      FOR j = 1 To 3
            READ grades(i, j)
            total(i) = total(i) + grades(i, j)
            PRINT grades(i, j);
      NEXT j
      ave(i) = total(i) / 3
      PRINT ave(i)
NEXT i
 Data A, 90, 80, 70
 Data B, 85, 75, 65
 Data C, 72, 44, 53
 Data D, 55, 66, 77
 Data E, 99, 99, 99
 Data F, 84, 85, 86
 Data G, 74, 75, 73
 Data H, 40, 45, 33
 Data K, 86, 83, 90
 Data L, 100, 87, 86
```

The arrays grade and ave could have been merged and made into one two-dimensional array. We purposely separated the two.

After the averages are calculated add the following code to sort the average in ascending order. Here we use a BASIC statement called **SWAP** if two values have to be interchanged.

SWAP *variable1, variable2*   Interchanges the values of *variable1* and *variable2*.

```
'Sorting the grades in ascending order
PRINT "The following grades are sorted in ascending order."
   FOR i = 1 TO 10
        FOR j = i + 1 To 10
           IF ave(i) > ave(j) THEN
              SWAP ave(i), ave(j)
              SWAP names$(i), names$(j)
              SWAP grades(i, 1), grades(j, 1)
              SWAP grades(i, 2), grades(j, 2)
              SWAP grades(i, 3), grades(j, 3)
           END IF
          NEXT j
```

```
    NEXT i
 FOR i = 1 To 10
 PRINT names$(i); grades(i, 1); grades(i, 2); grades(i, 3);
       ave(i)
 NEXT i
```

The program segment just above is a nested loop. Can you try and trace this program?

What will be the output of this program?

OUTPUT:

Objective E: An Interpretation of Matrix Multiplication (Optional)

Why is multiplication of matrices defined in this complicated way? One way to think about this is to see that matrices can be interpreted as a way of transforming one set of values into another set of values. Matrix multiplication then corresponds to doing one such transformation after the other.

For example, suppose matrix A has two rows and three columns, matrix B has three rows and two columns, and that both A and B (shown below) describe a mixing (transformation) process.

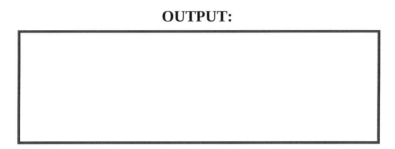

We start with two colors (call them a and b). Then the matrix B would transform the colors a and b as follows. It would take 1 part of a, and 2 parts of b, mix them together (add) to produce a new color we'll call c. This is just row 1 of B.

From what we've learned above, this is accomplished by simply taking the inner product of the first row of B with the column vector containing the colors a and b $\begin{bmatrix} a \\ b \end{bmatrix}$, i.e.,

$$[1\ 2] \bullet \begin{bmatrix} a \\ b \end{bmatrix} = (1*a + 2*b) = a + 2b = c$$

Continuing in this way, B would take 2 parts of a and 5 parts of b to mix a new color d. Which is just the inner product of the second row of B with the column vector of colors, i.e.,

$$[2\ 5] \bullet \begin{bmatrix} a \\ b \end{bmatrix} = (2*a + 5*b) = 2a + 5b = d$$

Finally, using the last row of B, we have that B takes 4 parts of a and 9 parts of b to yield a new color e, i.e.,

$$[4\ 9] \bullet \begin{bmatrix} a \\ b \end{bmatrix} = (4*a + 9*b) = 4a + 9b = e$$

Thus we say that B has transformed the colors a and b into three new colors c, d, and e.

Now we consider the 2 x 3 matrix A. A will mix (transform) the three new colors c, d, and e into two different colors f and g in the following way.

Matrix A would take 1 part of c, 2 parts of d and 3 parts of e, mix them together (add) to produce a new color we'll call f, i.e.

$$[1\ 2\ 3] \bullet \begin{bmatrix} c \\ d \\ e \end{bmatrix} = (1*c + 2*d + 3*e) = c + 2d + 3e = f$$

Finally, A would take 4 parts of c, 5 parts of d and 6 parts of e to create a new color g, i.e.,

$$[4\ 5\ 6] \bullet \begin{bmatrix} c \\ d \\ e \end{bmatrix} = (4*c + 5*d + 6*e) = 4c + 5d + 6e = g$$

If we recall that all these new colors are simply mixtures of the two original colors a and b, we see that:

$$f = c + 2d + 3e = (a + 2b) + 2(2a + 5b) + 3(4a + 9b) = 17a + 39b$$

$$g = 4c + 5d + 6e = 4(a + 2b) + 5(2a + 5b) + 6(4a + 9b) = 38a + 87b$$

Thus, color f is made up of 17 parts a and 39 parts b and color g has 38 parts a and 87 parts b.

Now notice that the matrix product AB (from above), yields

$$AB = \begin{bmatrix} 17 & 39 \\ 38 & 87 \end{bmatrix}$$

This is a new matrix that transforms our colors a and b into two new colors. The first color has 17 parts a and 39 parts b, while the second color has 38 parts a and 87 parts b. These contain the same ingredients as our colors f and g. Thus we see that matrix multiplication preserves the transformation (mixing) process.

This shows that by interpreting matrices as transformations then the matrix product is just a combination of two transformations, one after the other. It is this interpretation of matrix multiplication as the combination of two transformations that leads to the way matrix multiplication is defined.

Objective F: Inverse of a Matrix (Optional)

We have not mentioned anything about dividing matrices because there is no such thing as division of matrices. You can multiply matrices, but you cannot divide them. There is, however, a related concept called matrix inversion. Be forewarned, that matrix inversion is a very complicated operation that can be fairly difficult to understand the first time you encounter it!

Before we try to find the inverse of a matrix A, we must first know exactly what we mean by the (multiplicative) inverse. First we would like inversion of matrices to, in some way, be equivalent to division of real numbers.

For example, if we have an equation of the form: 2x=6 and we want to solve for x, we divide both sides by 2, which is equivalent to multiplying by ½ = 2^{-1}.

Thus we have:

$$2^{-1} 2\ x = 2^{-1}\ 6 \text{ or } x=6/2=3.$$

In general, the multiplicative inverse of a number "a," is the number, often written a^{-1}, with the property that

$$a^{-1} \times a = a \times a^{-1} = 1$$

For example, the inverse of 6 is the number 6^{-1} = 1/6, since (1/6) · 6 = 6· (1/6) = 1. (By the way, not every number has an inverse: the number 0 does not, and it is the only number without a multiplicative inverse.)

Now let's consider what we mean by the inverse of a matrix. First, note that when we say "inverse of a matrix" we mean the multiplicative inverse. Second, the inverse of a matrix will only be defined for square (n x n) matrices. The inverse of an n x n matrix **A** is defined to be that n x n matrix, which we write as $\mathbf{A}^{-1}$, that when multiplied by **A** on either side, yields the n x n identity matrix **I**. Thus,

$$\mathbf{AA}^{-1} = \mathbf{A}^{-1}\mathbf{A} = \mathbf{I}$$

If A has an inverse, it is called **invertible**. Otherwise, we say that it is **singular**.

Examples:

1. The inverse of the 1 × 1 matrix [3] is [1/3], since

$$[3][1/3] = [1] = [1/3][3]$$

2. Given the matrix $A = \begin{bmatrix} 1 & 1 \\ -2 & -1 \end{bmatrix}$, its inverse is the 2 × 2 matrix

$$A^{-1} = \begin{bmatrix} -1 & -1 \\ 2 & 1 \end{bmatrix}$$

The reader should multiply $\mathbf{AA}^{-1}$ as well as $\mathbf{A}^{-1}\mathbf{A}$ to see that both products give us

$$\begin{bmatrix} 1 & 0 \\ 0 & 1 \end{bmatrix}$$

3. The inverse of the n × n identity matrix **I** is **I** itself, since **I I = I**. So **I**$^{-1}$ = **I**

Given a n x n matrix, how do you find its inverse? The technique for inverting matrices is kind of clever. For a given matrix **A** and its inverse **A**$^{-1}$, we have

$$\mathbf{A^{-1}A = I}$$

We shall use the identity matrix **I** in the process of inverting a matrix.

The process we follow is as follows:
1. Form something we call an **augmented matrix** [**A**|**I**], where I is the n × n identity matrix and A is the n x n matrix we wish to invert.
2. Perform a series of valid "row operations" on the augmented matrix from above to get a matrix of the form [**I**|**B**].
3. Matrix B is the inverse matrix we are looking for.

To be able to carry this process out we need to introduce what we mean by valid "row operations" in item 2 above. A valid row operation on a matrix is one of the following procedures:
1. We can exchange any two rows in a matrix. We can multiply any row by a nonzero constant.
2. We can add a nonzero constant multiple of any row to a nonzero constant multiple of another row and replace the result with this new row.

The best way to introduce this concept is through an example. Consider the following matrix:

$$\mathbf{A} = \begin{bmatrix} 2 & 4 \\ 1 & -1 \end{bmatrix}$$

We wish to find **A**$^{-1}$ such that **A**$^{-1}$**A** = **AA**$^{-1}$ = **I**.

We follow the procedure above and form the augmented matrix:

$$\begin{bmatrix} 2 & 4 & | & 1 & 0 \\ 1 & -1 & | & 0 & 1 \end{bmatrix}$$

We now perform a series of row operations to transform this augmented matrix into the desired form. First we interchange the two rows to obtain:

$$\begin{bmatrix} 1 & -1 & | & 0 & 1 \\ 2 & 4 & | & 1 & 0 \end{bmatrix}$$

Next we add -2 times row 1 to row 2 and replace row 2 with the result:

$$\begin{bmatrix} 1 & -1 & | & 0 & 1 \\ 0 & 6 & | & 1 & -2 \end{bmatrix}$$

Next we multiply row 1 by 6 and add row 2 to it and replace row 1 with the result:

$$\begin{bmatrix} 6 & 0 & | & 1 & 4 \\ 0 & 6 & | & 1 & -2 \end{bmatrix}$$

Finally we divide each row by 6 to obtain:

$$\begin{bmatrix} 1 & 0 & | & 1/6 & 2/3 \\ 0 & 1 & | & 1/6 & -1/3 \end{bmatrix}$$

The inverse matrix is

$$A^{-1} = \begin{bmatrix} 1/6 & 2/3 \\ 1/6 & -1/3 \end{bmatrix}$$

The reader should verify that $A^{-1}A = AA^{-1} = I$.

COMPUTER ACTIVITY

Student Name_____ Section_____ Date_____

Identity Matrices (arrays)

1. Find the 2×2 identity matrix, then for matrix $C = \begin{bmatrix} 2 & 1 \\ 4 & 3 \end{bmatrix}$, show that

   IC = CI = C using the method shown is this section.

2. Find the 3×3 identity matrix, then for matrix $D = \begin{bmatrix} 2 & -1 & -2 \\ 4 & 1 & 9 \\ -3 & 0 & -5 \end{bmatrix}$, show that

   ID = DI = D

BASIC Programming

1. Write a program that returns the identity matrix with the same dimensions as matrix C above.

2. Write a program that returns the identity matrix with the same dimensions as matrix D above.

3. Modify the program to obtain the identity matrix for any n x n square matrix. You will need to use an **INPUT** statement that requests the order of the matrix for which you want to find the identity matrix.

COMPUTER ACTIVITY

Student Name_____ Section_____ Date_____

Scalar Multiplication

1. Given $A = \begin{bmatrix} 2 & 3 & -5 \\ 4 & 2 & 8 \end{bmatrix}$ and k = 6, Find 6**A**.

2. Information Technology Salaries (arrays)

The following figures represent the median hourly earnings excluding bonuses and benefits for IT occupations in three major cities as reported by Salary.com.

- An Applications System Analyst earns 24.78/hr, 23.97/hr, and 23.32/hr in New York, Los Angeles, and Chicago, respectively.

- A Software Engineer I earns 30.61/hr, 29.60/hr, 28.80/hr in New York, Los Angeles, and Chicago, respectively.

- A Help Desk Support I earns 21.38/hr, 20.68/hr, 20.12/hr in New York, Los Angeles and Chicago respectively.

We can summarize the above information in a table:

	New York	Los Angeles	Chicago
System Analyst	$24.78	$23.97	$23.32
Software Engineer	$30.61	$29.60	$28.80
Help Desk Support	$21.38	$20.68	$20.12

We can create a matrix with the above information using each row to represent an occupation and each column to represent a city.

NY   LA   Chicago

$$\mathbf{B} = \begin{bmatrix} 24.78 & 23.97 & 23.32 \\ 30.61 & 29.60 & 28.80 \\ 21.38 & 20.68 & 20.12 \end{bmatrix} \begin{matrix} \text{Systems Analyst} \\ \text{Software Engineer} \\ \text{Help Desk Support} \end{matrix}$$

To calculate the annual salary for each occupation, we multiply each hourly rate by 40, the number of working hours per week, and then multiply each result by 52, the number of weeks per year. We perform a scalar multiplication by multiplying each entry by 40*52 = 2080. Perform the scalar multiplication.

$$\mathbf{M} = (40*52)\mathbf{B} = (40*52) * \begin{bmatrix} 24.78 & 23.97 & 23.32 \\ 30.61 & 29.60 & 28.80 \\ 21.38 & 20.68 & 20.12 \end{bmatrix}$$

BASIC Programming Activity

1. Write a program to evaluate matrix **A** from above.

2. Write a program to evaluate matrix **M** from above.

3. Write a program that computes the annual salary for these occupations in each of the cities, and display the results in the form as indicated below.

	NY	LA	Chicago
Systems Analyst	?	?	?
Software Engineer	?	?	?
Help Desk Support	?	?	?

4. Modify the program using an **INPUT** statement and a **PRINT USING** statement to express the salary in $ xx,xxx.xx notation after computing the annual salary for any indicated hourly rate.

- What is the median annual salary for these occupations in each of the cities for a 35-hour work week and a 50-week work year?

5. Modify the program to calculate the annual salaries of these occupations in the three cities assuming the hours of the work week vary among the occupations.

- What are the median annual salaries of these occupations if the Applications System Analyst (ASA) works a median of 40 hours per week, the Software Engineer (SE) works 35 hours per week, and the Help Desk Support (HDS) works 50 hours per week?

EXERCISES 5.4

Let $\mathbf{A} = \begin{bmatrix} 2 & -3 & 4 \\ -1 & 2 & -5 \end{bmatrix}$, $\mathbf{B} = \begin{bmatrix} 1 & 0 & -3 \\ -2 & 1 & 3 \end{bmatrix}$, $\mathbf{C} = \begin{bmatrix} 1 & 0 & -1 \\ -3 & 2 & 4 \\ 0 & 7 & 0 \end{bmatrix}$, and $\mathbf{D} = \begin{bmatrix} 6 & -3 & 1 \\ -2 & 0 & -3 \\ 0 & -1 & 7 \end{bmatrix}$ be matrices. Perform the indicated operations below using the matrices given. If the operation is not possible, give the reason why.

1. **2A**
2. **−3C**
3. **A + B**
4. **A − 2B**
5. **A + C**
6. **−D**
7. **2C + D**
8. **D − B**
9. **AB**
10. **BD**
11. **CA**
12. **CD**
13. **BA**
14. $(2\mathbf{A}) \cdot \mathbf{b}$
15. **DB**
16. CD
17. DC

18. From your answers to the products **CD** and **DC**, explain why matrix multiplication is not commutative.

Section 5.5: Summation and Product Operators

Sometimes we need to add or multiply a long list of numbers. Rather than write out the entire list, mathematicians have created useful shorthand notations for representing these operations. The first is related to adding a list of numbers and is called the **summation operator**. The second is for multiplying a list of numbers, and this is called the **product operator**.

Objective A: Summation Operator

Suppose a child gets $1 allowance on the first day of the month, $2 on the second day, and so on, each day increasing by one dollar. How much money will the child have at the end of 30 days?

The solution to this allowance problem is $1 + 2 + 3 + 4 + \ldots + 29 + 30$. This is not difficult to add using a calculator, but suppose you want to know the sum at the end of the year, 365 days. This calculation can be tedious.

We shall introduce the summation notation and its components. We shall also show how we can write the repeated sum using this summation notation.

$$\sum_{k=1}^{n} a_k,$$

The Greek letter Σ (read sigma) is called a summation symbol. Sigma has the same sound as the English letter "s" (s for summation). The expression a_k can be a constant or an expression in terms of k. And the variable "k" is called the summation index. The index k keeps track of how many times addition has been done. Its initial value is 1, and it increments successively by 1 until the upper limit is reached. If we write the summation in expanded form, it looks like this:

$$\sum_{k=1}^{n} a_k = a_1 + a_2 + a_3 + \ldots + a_n,$$

Here, a_k is the expression being added, 1 is the lower limit and n is the upper limit.

A few notes about the summation index:
1. We have been using *k* as the summation index in the above discussions. We are not restricted to the letter *k*; we can use *i, j*, or any letter we choose.
2. The lower limit of summation index does not always have to start with 1. It could start with any number as long as it is lower than the upper limit. We shall show a couple of examples later.

Let us look at some examples of summation notation.

Example 1: Evaluate each of the following summations.

a) $\sum_{k=1}^{6} k$ b) $\sum_{k=1}^{4} k^2$ c) $\sum_{k=1}^{3} (2k+1)$

Solution: a) $\sum_{k=1}^{6} k = 1+2+3+4+5+6 = 21$ This is adding consecutive integers 1 to 6.

b) $\sum_{k=1}^{4} k^2 = 1^2 + 2^2 + 3^3 + 4^4 = 1+4+9+16 = 30$ This example adds the squares of 1, 2, 3, and 4.

c) $\sum_{k=1}^{3} (2k+1) = (2*1+1) + (2*2+1) + (2*3+1)$
$= 3+5+7 = 15$

This example adds only three numbers since *k* counts from 1 to 3. Each *k* is multiplied by 2 and then add 1 to the product.

Sometimes the lower limit is not 1, the summation notation looks like

$\sum_{k=m}^{n} a_k$, where $m \neq 1$ but $m \leq n$. This just means that we start our count of how many times addition has been done starting with whatever the lower limit is and continue until we reach the upper limit. Let us look at a couple of examples where the lower limit is not equal to 1.

Example 2: Evaluate each of the following summations.

a) $\sum_{k=4}^{7} k$ b) $\sum_{k=5}^{7} k^2$

Solution: a) $\sum_{k=4}^{7} k = 4+5+6+7 = 22$ We start with number 4, the lower limit.

b) $\sum_{k=5}^{7} k^2 = 5^2 + 6^2 + 7^2 = 25+36+49 = 110$

We shall now show how a simple formula can be "derived" to add consecutive integers and how to evaluate a summation where the expression a_k is equal to the summation index. We will then use it to solve our allowance example.

To derive the formual for adding consecutive integers

$$\sum_{k=1}^{n} k = 1+2+3+4+\ldots +(n-2)+(n-1)+n \qquad \text{Summation of k from 1 to n}$$

$$\sum_{k=1}^{n} k = n+(n-1)+(n-2)+\ldots +3+2+1 \qquad \text{Same sum written backwards}$$

$$2\sum_{k=1}^{n} k = \underbrace{(n+1)+(n+1)+(n+1)+\ldots +(n+1)+(n+1)}_{n \text{ of them}} \qquad \text{Adding corresponding Terms on both sides}$$

$$2\sum_{k=1}^{n} k = n(n+1) \qquad \text{Simplifying the right hand side of the equation equation}$$

$$\sum_{k=1}^{n} k = \frac{n(n+1)}{2} \qquad \text{This is the formula for adding consecutive integers}$$

Let us illustrate this formula with a simple example. Consider $1 + 2 + 3 + \ldots + 10$, we see that we are adding consecutive integers. We can write this summation as $\sum_{k=1}^{10} k$ and using the formula, we multiply the upper limit 10 by the next integer 11 which is 110 and then divide the product by 2 and we have 55.

$$\sum_{k=1}^{10} k = \frac{10(11)}{2} = \frac{110}{2} = 55.$$

Going back to the allowance example, it can be expressed in summation form. First note that $a_k = k$ since it is adding the consecutive integers. The lower limit is 1 and upper limit is 30. In summation notation, it is $\sum_{k=1}^{30} k$.

To evaluate this summation, notice that the upper limit is 30. Using the formula that we derived for adding consecutive integers, multiply 30 by the next integer which is 31 and then divide the product by 2, we have

$$\sum_{k=1}^{30} k = \frac{30(31)}{2} = 465$$

We see that the child will have $465 at the end of 30 days. You should try to calculate how much money the child will have at the end of 365 days.

If the expression a_k is a constant say 5, then $\sum_{k=1}^{n} 5 = 5+5+5+\text{L}+5 = n(5) = 5n$. Here, we are adding the number 5 n times, hence the result is 5*n or 5n.

Example 3: Evaluate $\sum_{k=1}^{4} 3$.

Solution: $\sum_{k=1}^{4} 3 = 3+3+3+3 = 4(3) = 12$

Suppose the expression is 2k, when we write the expanded form of the summation

$$\sum_{k=1}^{n} 2k = 2(1) + 2(2) + 2(3) + \text{L} + 2(n)$$
$$= 2(1+2+3+\text{L}+n)$$
$$= 2\sum_{k=1}^{n} k$$

The example above is just to show that the 2 in the expression is a common factor in each of the addends when the summation is expanded and so it can be factored. Then, notice that the expression inside the parentheses is the sum of consecutive integers.

There are some mathematical formulas that can be used to evaluate certain summation problems. We list them here for your reference.

1. $\sum_{k=1}^{n} k = \dfrac{n(n+1)}{2}$
2. $\sum_{k=1}^{n} ck = c\sum_{k=1}^{n} k = c\dfrac{n(n+1)}{2}$

3. $\sum_{k=1}^{n} c = nc$
4. $\sum_{k=1}^{n} k^2 = \dfrac{n(n+1)(2n+1)}{6}$

5. $\sum_{k=m}^{n} a_k = \sum_{k=1}^{n} a_k - \sum_{k=1}^{m-1} a_k$
6. $\sum_{k=1}^{n} (a_k + b_k) = \sum_{k=1}^{n} a_k + \sum_{k=1}^{n} b_k$

There are many more topics to discuss that are related to summation but they are beyond the scope of this text. The main focus of this section is to give an understanding of what the summation symbol means and be able to write a summation in expanded form.

For example, $\sum_{k=2}^{4} k^3 = 2^3 + 3^3 + 4^3 = 8 + 27 + 64 = 99$. Even though there is a formula that we can use, it is not hard to write out the three cubed numbers that will be added. Notice the lower limit is 2, so we start k with 2. We cube each of the numbers 2, 3, and 4, and add them.

Example 4: Evaluate the following: $\sum_{k=1}^{25} k$.

  Solution: Since this example is adding consecutive integers, we can use formula 1.

$$\sum_{k=1}^{25} k = \dfrac{25(26)}{2} = 325$$

Example 5: Evaluate the following: $\sum_{k=1}^{25} 3$.

  Solution: This example is adding the number 3 a total of 25 times, we use formula 3.

$$\sum_{k=1}^{25} 3 = 3 + 3 + 3 + \ldots + 3 = (25)3 = 75$$

Example 6: Evaluate the following: $\sum_{k=1}^{55} 3k$.

Solution: This example has a factor 3 in the expression, we use formula 2.

$$\sum_{k=1}^{55} 3k = 3\sum_{k=1}^{55} k = 3\left[\frac{55(56)}{2}\right] = 3(1540) = 4620$$

Example 7: Evaluate the following: $\sum_{k=15}^{55} k$.

Solution: This example does not begin with 1 as its lower limit, so we have to subtract. See formula 5.

$$\sum_{k=15}^{55} k = \sum_{k=1}^{55} k - \sum_{k=1}^{14} k = \frac{55(56)}{2} - \frac{14(15)}{2} = 1,435$$

To write a program segment to evaluate each of the above summations, we use a repetition structure or loop. Take the first example, $\sum_{k=1}^{25} k$, we can accomplish this by the following program segment:

```
SUM = 0
FOR K = 1 TO 25
     SUM = SUM + K
NEXT K
PRINT SUM
```

The output of this program segment should be 325. Try it and compare.

You should try to write programs for the other examples.

Objective B: Product Operator

Sometimes we have to multiply a list of numbers together. For example, if we want to multiply the first five counting numbers by themselves, i.e.,

$$1*2*3*4*5 = 120,$$

we can write this in a more condensed way. We introduce a new symbol to represent this, called the product operator, Π, and we can re-write the above multiplication as

$$1*2*3*4*5 = \prod_{k=1}^{5} k$$

Recall in chapter 4 we discussed the factorial notation, $5! = \prod_{k=1}^{5} k$.

Now, let us discuss a little more about this product notation. Π is the Greek letter pi, it has the same sound as the English "p" so we use this for the product notation.

Just like the summation we have a "product index" which keeps track of how many times the multiplication has been done. The expression can be equal to the index, an expression in terms of the index, or a constant. In short, the product operator does repeated multiplication instead of repeated addition. And just like the summation, what is being multiplied depends upon the expression.

Example 1: Evaluate $\prod_{k=4}^{8} k$

Solution: $\prod_{k=4}^{8} k = 4*5*6*7*8 = 6,720$

Example 2: Evaluate $\prod_{i=1}^{6} 2$.

Solution: $\prod_{i=1}^{6} 2 = 2*2*2*2*2*2 = 2^6 = 64$

Example 3: Evaluate $\prod_{k=2}^{5} (k+1)$.

Solution: $\prod_{k=2}^{5} (k+1) = (2+1)*(3+1)*(4+1)*(5+1) = 3*4*5*6 = 360$

Example 4: Evaluate. $\prod_{k=3}^{4} k^2$

Solution: $\prod_{k=3}^{4} k^2 = 3^2 * 4^2 = 9*16 = 144$

Example 5: Evaluate $\dfrac{\prod_{k=1}^{3} k}{\prod_{j=2}^{5} j}$.

Solution: $\dfrac{\prod_{k=1}^{3} k}{\prod_{j=2}^{5} j} = \dfrac{1*2*3}{2*3*4*5} = \dfrac{1}{20}$

How do we write the program segment that would evaluate these products? It is still a loop, however, we are repeatedly multiplying rather than adding.

Example 6: Write the program segment to evaluate $1*2*3*4*5 = \prod_{i=1}^{5} i$:

```
PROD = 1
FOR K = 1 TO 5
     PROD = PROD*K
NEXT K
PRINT PROD
```

The output for this program segment should be 120.

Example 7: Write the program segment to evaluate $\prod_{k=2}^{5} (k+1)$:

```
PROD = 1
FOR K = 2 TO 5
     PROD = PROD*(K + 1)
NEXT K
PRINT PROD
```

The output for this program segment should be 360.

Notice that PROD is initialized to 1 instead of 0 as in summation. This is because if it was initialized to 0, then PROD will always be 0.

Try to modify the program segment above to solve the other examples.

EXERCISES 5.5

Evaluate each by expanding or using the properties of summation notation.

1. $\sum_{i=1}^{7000} i$

2. $\sum_{j=1}^{94} 9$

3. $\sum_{i=1}^{482} i$

4. $\sum_{i=800}^{7000} i$

5. $\sum_{k=6}^{60} 17$

6. $\sum_{k=1}^{20} (k+4)$

7. $\sum_{i=1}^{4} (\frac{3}{5}i + 7)$

8. $\sum_{j=10}^{55} j$

9. $\sum_{k=2}^{6} (3k^2 - 8)$

10. $\sum_{j=5}^{7} j^3$

11. $\sum_{i=1}^{5} 2i^2$

12. $\sum_{k=1}^{15} (2k+5)$

Evaluate each of the following products.

13. $\prod_{i=1}^{6} i$

14. $\prod_{i=1}^{7} 4$

15. $\prod_{k=3}^{7} 3k$

16. $\prod_{k=1}^{4} \frac{3}{5}k$

17. $\prod_{j=3}^{8} \frac{j}{j+1}$

18. $\prod_{k=1}^{4} k^2$

19. $\dfrac{\prod_{i=2}^{6} i}{\prod_{j=3}^{5} j^2}$

20. $\prod_{j=1}^{5} \frac{j+2}{j}$

Chapter 6

Set Theory – Preliminary Concepts

The collection and the representation of objects or data is an important study in mathematics and computing. Organizing, accessing, and changing data are functions performed by employees of the Information Technology areas routinely. Performing these tasks can be time consuming as well as error prone since the vast amount of data one needs to shuffle through can range in the millions and billions of records.

Consider a company that keeps records of its 5000 employees that include: employee last name, first name, employee I.D, date hired, date of birth, social security number, salary, vacation time allotted, sick days allotted, personal days allotted, city of residence, etc. You should sense that this could amount to a lot of information. The company may want to compile a report that gives them a list with the certain criteria they need. Suppose for example, the company wants to reward a special type of employee with a bonus based on their performance and dedication to the company. They may need to generate a report listing those employees that fit specific criterion. A query may be initiated for a record of employees that are in management positions that have worked for the company for more than five years or have not taken any vacation time in the last three years, and are making more than $100,000 per year salary.

A relational database linking tables that contain this type of data may be useful. As it turns out, a relational database is a nice application of where sets arise in the computing world. A thorough discussion on databases and relational databases is not included in this text. A textbook or course on databases will provide moredetailed discussion on these concepts. We encourage those interested in information technology to gain some knowledge in this area. Set theory provides the fundamental building blocks of databases and that is what this chapter will focus on.

In this chapter, we will discuss properties associated with sets and discuss the connection that sets have in programming and computing.

After completing the chapter, you should be able to :
- read a set and identify the objects in that set,
- create new sets by applying operations such as union, complement, and intersection,
- compare sets and determine whether two sets are equal or equivalent,
- illustrate sets with Venn diagrams,

- understand the connection between sets and logic that we studied in Chapter 3, and
- use concepts learned about Sets and apply them to programming and computing.

Section 6.1: Sets and Set Notation

Objective A: Describing Sets

Each of the ideas discussed in this chapter have a common theme; to describe the members that belong to a set either explicitly, one by one, or with a description, thus allowing the person reading the set to understand what belongs to the collection.
We begin this section with some of the basic terminology, notation, and descriptions of sets.

Definition: A **set** is a well defined collection of objects called elements. These elements are often referred to as members of the set.

It is customary to call or name a set by a capital letter such as A, B, C and the like. We enclose the members by braces { }.

We will discuss the meaning of the term well defined later in this section. For now, however, you take it to mean that the members of the set are clearly described and not ambiguous.

Example 1: List the elements of the set:
   S={a, b, c, d }

   Solution: In this example, the set name is S and the members or elements of the set are the letters a, b, c, and d.

Example 2: List all the members of the set A={1, 2, 3, …10}
   Solution: The set above is a set that is named A and the elements of the set are all the integers from 1 through 10.   i.e. $A = \{1, 2, 3, 4, 5, 6, 7, 8, 9, 10\}$

Remember from Chapter 1, three dots … called ellipses indicate a continuation of the same pattern and its notation commonly used in the discussion on sets.

There are three ways of describing sets we will use. We will learn to define sets by giving a description of a set in English, use roster form to define a set, and use mathematical set-builder notation to define sets.

If we describe a set, the description must be clear enough that the reader so that the reader knows what elements are in the set. If the elements are listed the set is said to be in *roster* form.

Consider the following examples:

Example 3: List the elements of the set A = The days of the week ,in roster form.

  Solution: We use the description to represent the set in roster form. The set is:
      A={Monday, Tuesday, Wednesday, Thursday, Friday, Saturday, Sunday}.

Example 4: The set A is the set of all digits in the binary number system. List set A in roster form.

  Solution: Since 1 and 0 are the only digits of the binary number system, A = {0,1}

We commonly use sets to define the Real number system. The letters assigned to those sets are standard and should be recognized when referred to. The sets below have been define in Chapter 1, but we list them below again for your reference.

The set of <u>Natural numbers</u> is denoted by the letter **N** where
  N={1, 2, 3, 4, 5, …}

The set of <u>Whole numbers</u> is denoted by **W** where
  W={0, 1, 2, 3, 4, …}.

The set of <u>Integers</u> is denoted by **Z** where
  Z={…-4, -3, -2, -1, 0, 1, 2, 3, 4, …}

Note: In order to properly define Z, it was necessary to use the ellipses in both directions. Although the list of elements in these sets is infinitely long, the sets are in listed in roster form.

Set-Builder Notation:

Another method used to describe sets is set-builder notation. Set-builder notation uses mathematical notation to describe sets without having to rely on too many words in the description.

An example of this is the set of rational numbers. We can describe the set of rationals as the collection of ratios such that the numerator and denominator of each ratio is a member of the integers provided that the integer in the denominator is not 0
The standard form of using set builder notation is:

$S = \{x \mid \text{condition or description for } x\}$. |This is read as " the set S is the collection of all x "such that" the condition is satisfied by x". The vertical bar in this notation means "such that".

The set of rationals is denoted by **Q**, for quotients, where

$\mathbf{Q} = \{\frac{p}{q} \mid p, q \text{ is a member of } \mathbf{Z} \,;\, q \neq 0\}$. This is read as the set **Q** is the collection of all possible fractions $\frac{p}{q}$, such that p and q can be any elements from the set of integers, but q cannot be 0 since we can never divide a number by 0.

This method for representing sets is extremely useful here since it is difficult to list "all" the rational numbers.

From this definition, we can see that the fraction or rational number $\frac{7}{11}$ is described in the set above by the integers $p = 7$ and $q = 5$.

Example 4: Describe the elements of the set $A = \{x \mid x \text{ is a day of the week}\}$.

Solution: The elements of the set A are Monday, Tuesday, Wednesday, Thursday, Friday, Saturday and Sunday.

Example 5: Describe the elements of the set $B = \{x \mid x^2 = 36\}$.

Solution: This is the set of all numbers x such that the square of x is 36, so the members of B are 6 or –6. In roster form, $B = \{-6, 6\}$.

It is often helpful to indicate membership of an element in a set by the symbol $\in$ which means " belongs to" or " is a member of". In our above example we can say that $6 \in B$ and $-6 \in B$. The symbol $\notin$ means "does not belong to" and we can use it to say $5 \notin B$.

Example 6: Describe the elements of the set $C = \{x \mid x \in N \text{ and } x^2 = 81\}$.

Solution: The solution to the equation $x^2 = 81$ is 9 or –9. Since –9 is not a natural number, it cannot belong to set C. Therefore C={9}. Note that C has only one member.

Example 7: Describe the elements of the set $A = \{x \mid x \in N \text{ and } x \geq 10 \text{ and } x < 15\}$ in roster form.

Solution: The elements of the set can be described in roster form by
$A = \{10, 11, 12, 13, 14\}$.

In example 7 above, one of the conditions stated is that x must be a natural number. It is sometimes more convenient for notational purposes only to state the set above as $A = \{x \in N \mid x \geq 10 \text{ and } x < 15\}$. This states that the elements that satisfies the given statement must be integers and greater than or equal to 10, but less than 15.

It is also possible for sets to be members of a set. For example the set A={1, 2, {3}, 4} is the set whose members are 1, 2, {3}, and 4. Note that 3 is not an element of set A, however, the set that contains the element 3 is member of set A.

Example 8: Use $\in$ or $\notin$ wherever appropriate.
    a) 8 _____ {4,5,6,7,8}
    b) 24 _____ {2,4,6,8,10,...30}
    c) b _____ {m,a,t,h}
    d) 2 _____ {1,2,3,4}
    e) {2} _____ {1,2,3,4}

Solution:
    a). Since 8 is an element of the set, we can write $8 \in \{4,5,6,7,8\}$.
    b) If we continue this pattern of even numbers, it is clear that 24 is a member of the set. We can therefore write $24 \in \{2,4,6,8,10,...30\}$
    c) The element b does not belong to the set. We can then write
        $b \notin \{m,a,t,h\}$
    d) Since 2 is a member of the set, we can write $2 \in \{1,2,3,4\}$.
    e) Although the number 2 is an element, the set {2} is not a member of the set. Therefore we write $\{2\} \notin \{1,2,3,4\}$

Whenever we define sets, the description must be clear enough and contain all the information necessary to allow the reader to determine what elements are in the set.

Consider for example a set of the form C = {all good video games}. The description of C is vague since we do not know what is considered a "good" video game are. We would need more information to determine what elements are in set C.

How well a set is defined is of great importance in determining the members of the set. This notion gives rise to the notion of a well-defined collection defined below.

Definition: A **well-defined** collection is a set whose members can be completely determined by the given conditions.

The definition means that the description of the sets must be clear enough so the one reading the description can determine.

Each of the examples of sets we have encountered thus far were well-defined. To explore some that are not well defined, we consider the next few examples:

Examples: Explain why the collections are not well-defined
9. $S = \{x \mid x$ is a month with average daily temperature greater than 75 degrees$\}$
10. The set of cool games on the internet
11. $A = \{x \mid$ a good computer$\}$

Solutions:
9. The collection S is not a well defined set since it is unclear which months belong to the set. Temperature can vary from region to region.

10. The word "cool" has different meanings to different people and we cannot clearly determine which games belong to the set. The set is not well defined.

11. This set is also not well defined. Clearly what one thinks to be a good computer, another may think is a lousy one.

Objective B: The Universal Set and the Empty Set

We now introduce the universal and the empty set.

Definition: The **universal set**, usually denoted by U, is the set that contains all the possible elements in a discussion.

Example 1: $U = \{x \in W \mid x \leq 10\}$.

Solution: It is clear that the Universal Set contains all the whole numbers that are less than or equal to 10. i.e. $U = \{0, 1, 2, 3, 4, 5, 6, 7, 8, 9, 10\}$

Example 2: What might be the universal set in the discussion of the set of female students taking a mathematics course for information technology during the fall 2004 semester at Mathematica University.

Solution: Since the set consists only of the female students taking information technology mathematics during the fall 2004 semester at Mathematica University during the specified time, the universal set might be the set of all (male and female) students taking this course during the fall 2004semester. The female students are then a part of this universal set.

It is not necessary for every set to have members or elements that belong to it. Consider the set $S = \{x \in N \mid x^2 < 0\}$. S is the set that contains all the natural numbers whose squares are less than 0, i.e. negative. We know that the square of every number is non-negative. The set S therefore has no elements and in roster form is described by $S = \{\ \}$. Another notation to indicate a set with no elements is $\emptyset$. The following definition clarifies the preceding statements.

Definition: The **empty set** or **null set**, denoted by $\{\ \}$ or $\emptyset$, is the set that contains no elements.

Example 3: Describe the elements of the set $A = \{x \mid x^2 = -81\}$.

Solution: The set A is the null or empty set. It is impossible to find a member of A since it is impossible to square any real number and get a negative number as a result. We say that $A = \{\ \}$ or $A = \emptyset$.

Example 4: Describe the elements of the set
$S = \{$ The set of digits greater than one in the binary number system$\}$

Solution: Since the only digits in the binary number system are 0 and 1, there are no digits greater than 1. The set therefore has no members and we say that $S = \{\ \}$ or $S = \emptyset$.

A common mistake is to view the set $\{\emptyset\}$ as an empty set when in fact it isn't. The set $\{\emptyset\}$ is the set that contains the empty set. It is therefore not empty.

EXERCISES 6.1

Use set notation to list all the elements of the set:
1. S is the set of all counting numbers less than 10.
2. The set of all integers between –3 and 5 inclusive.
3. $A = \{-2, 0, 2, 4, \ldots 16\}$
4. $\{x \mid x \text{ is an even integer between 14 and 30}\}$
5. $\{x \mid x \text{ is an odd integer between 7 and 21 inclusive}\}$
6. $\{x \mid x \in N \text{ and } x^2 = 144\}$
7. $\{x \mid x^2 = -36\}$
8. $\{x \mid x \in Z \text{ and } x^2 - 1 = 24\}$
9. $\{x \mid x \in N \text{ and } -2 \leq x < 7$
10. $\{x \mid x \in Z \text{ and } -3 < x < 11$

Denote each by set builder notation.
11. $\{0, 5, 10, 15, 20\}$
12. $\{2, 4, 6, 8, \ldots, 30\}$
13. The set of all even natural numbers.
14. The set of all odd natural numbers
15. The set of all even integers

Use $\in$ or $\notin$ wherever appropriate
16. $2 \_\_ \{0, 1, 2, 3, 4, 5\}$
17. $\{2\} \_\_ \{0, 1, 2, 3, 4, 5\}$
18. $a \_\_ \{m, a, t, h\}$
19. $\{-2\} \_\_ \{-3, -2, -1, 0, 1, 2, 3, 4\}$
20. $\{b\} \_\_ \{a, \{b\}, \{c\}, d, e\}$
21. $\{1, 2, 3\} \_\_ \{1, 2, 3, 4\}$
22. $2 \_\_ \emptyset$
23. $\sqrt{2} \_\_ W$
24. $\sqrt{2} \_\_ Q$
25. $\sqrt{4} \_\_ Z$

Determine whether the sets are well defined.

26. The set of all people in NYC on January 1, 2004.
27. The set of all big buildings on Manhattan Island
28. $\{x \mid x \text{ is a fun course}\}$
29. $\{x \mid x \text{ is a digit in the hexadecimal number system}\}$
30. The set of brand name computers with fast processors.

Section 6.2 Subsets and Proper Subsets

Objective A: Subsets

It is possible to have two sets that share none, some, or all elements. For example, If the set A={1,2,3,4,...10} and set B={2,4,6,8,10}, then every elements of B is also an element of A. When two sets have this special type of relationship, we say that B is a subset of A. We will use Venn diagrams to illustrate subsets. The rectangular region of the diagram contains all the elements of A and the circular region inside is a subset of A.

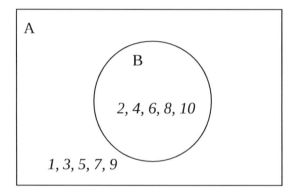

The elements contained in B are elements that also belong to set A. however, the elements 1,3,5,7,9, belong to set A but not to set B.

Definition: Set B is said to be a **subset** of A, denoted by $B \subseteq A$, if every element of set B is also an element of set A.

Another example if $A = \{1,2,3\}$, $B = \{1\}$, and $C = \{1,2,5\}$, we see that B is a subset of A, B is a subset of C, but A is not a subset of C. In addition, set A is a subset of itself since every element of A is an element of itself. The same can be said about sets B and C.

Note that $\emptyset \subseteq A$. In fact, the empty set is a subset of every set and every finite set is a subset of itself. We will discuss these concepts in more detail throughout the section.

If a set A is not a subset of a set B, then we denote this by: $A \not\subseteq B$.

Now that we have define the various sets of the real number system such as **N, Z, W, Q**, and **I**, we can state a partial list of subsets such as:

1. $N \subseteq W$
2. $N \subseteq Z$
3. $W \subseteq Z$
4. $N \subseteq Q$

Example 1: Determine whether $\{red, blue, green\} \subseteq \{x \mid x \text{ is a color}\}$ is True.

Solution: This is true since $\{x \mid \text{ is a color}\}$, contains a list of all colors, we can say $\{red, blue, green\} \subseteq \{x \mid x \text{ is a color}\}$ are a color.

Example 2: Determine whether $\{x \mid x \text{ is a color}\} \subseteq \{red, blue, green\}$ is True.

Solution: This is not true since orange is a member of $\{x \mid x \text{ is a color}\}$, but is not a member of $\{red, blue, green\}$. We then say
$\{x \mid x \text{ is a color}\} \not\subseteq \{red, blue, green\}$

Given any non-empty set, we can find all subsets of that set by the procedure outlined in the next example.

Example 3: Find all subsets of the set $A = \{a, b, c\}$

Solution: To find the subsets of set A, we will list the sets in categories where we first list the empty set, then the subsets of A that contain only one element, followed by the subsets of A that contain two elements, and so forth until the last category lists the set itself.

The subset of set A above are:
∅   {a}   {a,b}   {a,b,c}
    {b}   {a,c}
    {c}   {b,c}

We see that set A has 8 different subsets which is 2^3.

Example 4: List all subset of the set $S = \{\text{dog, cat}\}$

Solution: We proceed just as we did in the previous example but we now have only two elements in the set. The subsets are:
∅   {dog}   {cat}   {dog,cat}

There are only 4 subsets of a set with 2 elements which is 2^2.

In fact any set with n elements will have 2^n subsets.

Objective B: Proper Subsets

The proper subsets of a set A are similar to the subsets of set A except that the set itself is not considered in the collection.

Definition: Set A is a **proper subset** of a set B, denoted by $A \subset B$ if:
1. $A \subseteq B$ (A is a subset of B) and
2. There is at least one element that is in B but NOT in A.

Example 1: Determine whether $\{2,3,4\} \subset \{1,2,3,4\}$.

Solution: Since $\{1,2,3,4\}$ has an element (1) that is not in $\{2,3,4\}$ and there is at least one element of $\{1,2,3,4\}$ that is not in $\{2,3,4\}$, we say that
$\{2,3,4\} \subset \{1,2,3,4\}$

Example 2: Determine whether $\{\ \} \subset \{a,b,c,d\}$.

Solution: Since $\{a,b,c,d\}$ has elements that are not in $\{\}$ we say that $\{\} \subset \{a,b,c,d\}$ is true.

In fact, based on example 2 above, we say that the empty set is a proper subset of any non-empty set. Furthermore, the empty set is a subset any set.

Now let's discuss how we list all proper subsets of a set. We will illustrate this by example.

Example 3: List all the proper subsets of set $A = \{a,b,c\}$

Solution: The proper subsets are:
∅, {a}   {a,b}
     {b}   {a,c}
     {c}   {b,c}

Note that proper subsets are constructed just as the subsets are constructed with the exception that the set itself is not included.

In summary, the preceding concepts lead to some interesting facts we should be aware of. We list those facts here for reference.

1. Every set is a subset of itself
2. No set is a proper subset of itself.
3. A set with n element has 2^n subsets
4. A set with n elements has $2^n - 1$ proper subsets.

Observe that the set A in example 3 has 3 elements and has $2^3 - 1 = 7$ proper subsets.

Example 4: List all the proper subsets of set $S = \{0,1\}$

   Solution: Since S has two elements there are $2^2 - 1 = 3$ proper subsets. They are $\{\},\{0\},\{1\}$.

Venn Diagram Illustrations of Subsets and Proper Subsets:

We introduced Venn diagrams earlier when we introduced subsets. Venn diagrams provide a pictorial representation of sets and subsets. From now on, the inside of the rectangular region will contain all the elements that belong to the universal set, and the sets inside this region "group" those elements when appropriate to the problem.

The illustration below indicates that $A \subseteq U$ (A is a subset of U). Since U does not contain an element that is not in A, A is *not* a proper subset of U and so we can write $A \not\subset U$.

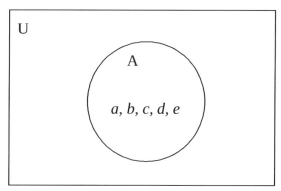

The illustration below shows that $A \subseteq U$. Furthermore since U contains at least one element that is not in A, we can further say that A is a proper subset of U, or $A \subset U$.

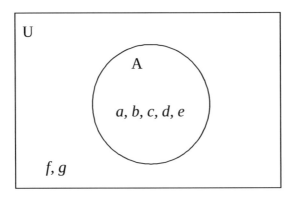

In the next diagram B={1,2,3} and C = {1,2,3,4}. We see that B is a subset of C since each element in B is also an element in C. In fact, B is not just a subset of C, but also a proper subset of C. Also note that C contains every element inside of its circular picture.

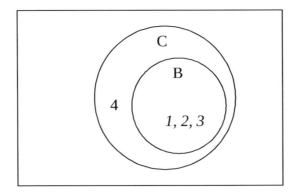

Objective C: Complement Operation of Sets

It is sometimes useful to describe elements that are in the universal set but not in a subset of that universal set. The illustration below shows set A that is a subset of U. The elements of U that are not elements of A are said to belong to the complement of A. The example below illustrates that $U = \{Bill, John, Jack, Mike, Mary\}$, and $A = \{John, Mary\}$. Clearly Jack, Mike, and Bill do not belong to set A. We will say that they belong to the complement of A.

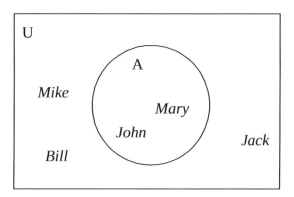

The complement of A is also considered a set and its precise definition and notation is given below.

Definition: The **complement** of a set A, is denoted and defined by
$\overline{A} = \{x \mid x \in U \text{ and } x \notin A\}$, i.e. $\overline{A}$ is the set that contains all the elements in the universal set that are not in set A.

Other notation commonly used to denote the complement of a set such include A' or A^c, but in this text we will use $\overline{A}$.

Example 1: Find the complement of set A if:
$$U = \{a,b,c,d,e,f,g,h\}$$
$$A = \{a,d,e,g\}$$

Solution: The complement of the set contains all the members of the universal set that are not in set A, therefore $\overline{A} = \{b,c,f,h\}$

Example 2: Find $\overline{A}$, $\overline{B}$, and $\overline{U}$ given the sets
$$U = \{m,a,t,h\}$$
$$A = \{m,h\}$$
$$B = \{t,h\}$$

Solution: The sets $\overline{A}, \overline{B}, \overline{U}$ contain all the elements in the universal set that are not in set A, B, or U respectively. Therefore $\overline{A} = \{a,t\}$ and $\overline{B} = \{m,a\}$.

Since the universal set contains all the elements in the discussion, the complement of U must not contain any of those elements. The only set that makes sense in this case is the empty set. Therefore, we can say
$$\overline{U} = \{\} \text{ or } \overline{U} = \emptyset.$$

The following are always true for any universal set U and empty set $\emptyset$

  a. $\overline{U} = \emptyset$
  b. $\overline{\emptyset} = U$

We have seen so far that when we take the complement of a set, it results in another set. For example, the complement of set A is the set $\overline{A}$. We can further take the complement of the complement, that results in another set denoted by $\overline{\overline{A}}$, the *double* complement, Lets take a closer look at $\overline{\overline{A}}$.

We know that if an element belongs to a set A then it does not belong to the set $\overline{A}$. Furthermore if an element belongs to the set $\overline{A}$, then it must not belong to the set A. We can summarize this by the statement:

If $A \subseteq U$ and $x \in U$, then either $x \in A$ or $x \in \overline{A}$ but not simultaneously to both.

Mathematically, we can argue that if

$x \in \overline{\overline{A}}.$

then $x \notin \overline{A}$

therefore it must be that $x \in A$

An example is given below that outlines the *double* complement. This is an important fact that will be used later in the course since the double complement of a set resulting in the set itself.

Example 3: List the elements of the set $\overline{\overline{A}}$ given :
$U = \{1, 2, 3, a, b, c, d\}$
$A = \{1, b, d\}$

Solution: The set $\overline{\overline{A}}$ denotes the double complement of set A. We begin by:

$$A = \{1, b, d\}$$
$$\overline{A} = \{2, 3, a, c\} \text{ then so is } \overline{\overline{A}}.$$
$$\overline{\overline{A}} = \{1, b, d\}$$
That is $\overline{\overline{A}} = \{1, b, d\}$

Example 4: Given the sets below, find $\overline{\overline{U}}$
$$U = \{l, o, g, i, c\}$$
$$A = \{i, c\}$$

Solution: The elements of $\overline{\overline{U}}$ are the same as the elements of U. Therefore,
$$\overline{\overline{U}} = U = \{l, o, g, i, c\}$$

EXERCISES 6.2

Use $\subseteq, \subset, \not\subseteq$ or say neither wherever appropriate. Enter all that apply.

1. $\{1, 2, 5\}$ ___ $\{1, 2, 3, 4, 5, 6\}$
2. $\{1, 3, 5, 7, 9\}$ ___ $\{3, 5, 7\}$
3. $\emptyset$ ___ $\{x \mid x \text{ is a warm state on the east coast of America}\}$
4. 2 ___ $\{1, 2, 3, 4, , 6\}$
5. $\{\text{math, physics, english}\}$ ___ $\{x \mid x \text{ is a college discipline}\}$
6. car ___ $\{\text{boat, car, helicopter}\}$
7. $\{-4, 4\}$ ___ $\{x \mid x \in Z \text{ and } x^2 = 16\}$
8. $\emptyset$ ___ $\{x \mid x \in N \text{ and } x < 12\}$
9. $\emptyset$ ___ $\{\ \}$
10. $\{\ \}$ ___ $\{\emptyset\}$

Find the indicated set if $U = \{a, b, c, d, e, f, g\}$ *is the universal set of each of the sets below*
11. Given $A = \{e, f, g\}$ find $\overline{A}$.
12. Given $C = \{a, b, c\}$ find $\overline{C}$.
13. Given $S = \{a, c, e\}$ find $\overline{\overline{S}}$.

14. Given $B = \emptyset$, find $\overline{B}$

List all the subsets of the given set by first determining how many there are.

15. $S = \{cat, dog\}$
16. $A = \{1, 2, 3\}$
17. $B = \{x \mid x \in N \text{ and } x-7=12\}$
18. $\{x \mid x \text{ is a binary digit}\}$
19. $C = \{a, b, c, d\}$
20. $P = \{x \mid x \in N \text{ and } x \leq 0\}$

22-26 List all the proper subsets of the sets above, (16-21), by first determining the number of proper subsets for each set. If there are no proper subsets, then say so.

27. Illustrate the following by a Venn diagram.
    U={1,2,3,4,5,6,7,8,9,10}
    A={2,4,5,9}

28. Illustrate the following by a Venn diagram, where U denotes the universal set.

    U={1,2,3,4,5,6,7,8,9,A,B,C,D,E,F}
    S={2,4,5,A,D}
    T={2,4,5,A,D,F}

29. Illustrate by Venn diagram the following subsets of the real number system:
    a) $N \subset W$
    b) $W \subset Z$
    b) $Z \subset Q$
    c) $Q \subset R$
    d) $N \subset W \subset Z \subset Q \subset R$

Section 6.3: Comparison and Cardinality of Sets

In this section we will explore two different ways of comparing sets. These two notions entail that of equal sets or equivalent sets. These two notions are different. Our intuition tells us that two things are equal if they are identically the same. In set theory, we are also interested in determining when two sets are equal. As the definition will illustrate, the necessary condition for the sets to be equal is similar to being identical. There is however the term equivalence we will also discuss. Equivalent sets do not guarantee they are identical.

Objective A: Set Equality

Earlier in the chapter, we discussed that we never duplicate elements in a set. Remember that if a set contains all elements of a particular discussion and if the elements happen to appear more than once in that discussion, it is not necessary to re-list them. Consider the set $A = \{x \mid x = (-1)^n,\ n \in N\}$. The elements of the set are determined by:

$$(-1)^1 = -1$$
$$(-1)^2 = 1$$
$$(-1)^3 = -1$$
$$(-1)^4 = 1$$

The only elements of A are 1 and –1. The fact is that this occurs without end, but since these are the only elements; $A = \{-1, 1\}$. We say that set A has two elements.

As another example, the set $B = \{x \mid x^2 = 1\}$ also has two elements. The solutions to the equation $x^2 = 1$ are –1 and 1 and so the elements that are in set B are indicated by $B = \{-1, 1\}$.

You should notice that set A and set B are determined by completely different rules but they have the exact same elements. This is common in mathematics.

Definition: Two sets A, B are said to be **equal**, denoted by A = B, if they each contain the exact same elements.

The sets A and B from the previous example each have the same elements. Therefore they are said to be equal and we write A = B.

Example 1: Determine which of the sets below are equal.
$$A = \{Joe, Harry, Tina\}$$
$$B = \{George, Tina, Harry\}$$
$$C = \{Tina, Harry, Joe\}$$
$$D = \{George, Tina, Harry, Maria\}$$

Solution: The order in which the elements appear is not important. What is important is whether the sets contain the same elements. Only sets A and C contain the same elements. Therefore $A = C$.

Objective B: Cardinality of Sets

Counting the elements of a set is also common practice and very useful to us. We refer to the cardinality of the set as being the number of elements the set contains. Cardinality is in essence the "size" of the set.

Definition: The **cardinality** of a set S, denoted by $n(S)$ is the number of elements in the set.

Some textbooks denote the cardinality of a set S by $|S|$. In this text, we will use $n(S)$.

We can consider some of the previous sets we used and find their cardinalities.

Examples: Find the cardinality of the sets below:

1. $A = \{1, 2, 3, a, b\}$
   $B = \{a, b, d, 1, 2, 3\}$

2. $C = \{a, b, c, d\}$
   $D = \{1, a, 2, g\}$

3. $S = \{x \mid x \in W, \ 0 \leq x \leq 14\}$

Solutions:

1. Since set A has five elements, $n(A) = 5$, and since set B has six elements, $n(B) = 6$.

2. Since set C has four elements, $n(C) = 4$, and since set D has four elements, $n(D) = 4$.

3. The elements of the set are $S = \{0, 2, 4, 6, 8, 10, 12, 14\}$. There are 8 elements in the set, therefore $n(S) = 8$

Objective C: Set Equivalence

There is yet another notion that relates one set to another set. The concept is that of *equivalence*. When determining whether two set are equivalent, we do not look at the individual members, but rather at the cardinality of the sets. We are interested in whether the two sets have the same *number* of elements. As discussed earlier, when the two sets have the same number of elements and the elements also happen to be identical, we will call them equal sets, however, if the elements do not happen to be the same, we will call them equivalent sets.

Definition: Two sets A, B are said to be **equivalent**, denoted by $A \approx B$ if they each have the same number of elements. e.g. $A \approx B$ if and only if n(A)=n(B).

This means that regardless of what the elements are, we are only concerned with the cardinality of the two sets.

Example 1: Determine whether the sets are equivalent, equal, or neither:
$C = \{a, b, c, d\}$
$D = \{1, a, 2, g\}$

Solution: Sets C and D clearly do not have the exact same elements; hence they are not equal. However, since $n(C) = 4$ and $n(D) = 4$, set A is equivalent to set B, and hence, $A \approx B$.

Example 2: Determine whether the sets are equal, equivalent, or neither.
$A = \{1, 2, 3, a, b\}$
$B = \{a, b, d, 1, 2, 3\}$

Solution: The sets are clearly not equal since they do not contain the exact same elements, therefore $A \neq B$. In addition, since $n(A) = 5$ and $n(B) = 6$, the sets are also not equivalent. The answer is therefore *neither*.

One should note that if a set A is equal to a set B, they are also by default equivalent. Certainly having the exact same elements implies that they each have the exact same *number* of elements.

It is prudent to mention that although the above statement is accurate for this text, the statement must be modified when discussing cardinalities of infinite sets. In the next objective, we will give an overview of infinite sets without discussing the cardinality of infinite sets. For those interested, a set theory textbook will offer a detailed discussion or an internet search of cardinality of infinite sets will return many college and professional websites that explain the concept thoroughly.

Objective D: Finite and Infinite Sets

Collections of objects can be classified as finite collections or infinite collections. The difference in these concepts lies primarily on whether you can count the elements in the collection or cannot count the elements in the collection. For us to be able to count the elements in a collection, it will be necessary for the set to have a beginning element and a final element.

Suppose we look at the set $S = \{1, 3, 5, 7, 9, ..., 21\}$. This set contains the odd counting numbers between 1 and 21 inclusive. It has a beginning element as well as an ending element. Sets such as this are said to be finite.

On the other hand, suppose you generate a list of real numbers that lie between the number 0 and 1. One can begin writing a few of the numbers that belong to the list as:

    .1
    .01
    .001
    .0001
    .00001
    .000001

and so forth. You should get the feel that this list of numbers is never-ending. The number .0000…1 is part of that list regardless of the number of decimal places the number has.
The set could be stated as $\{.1, .01, .001, .0001, .00001, .000001 \ldots\}$

Sets such as this are said to be infinite sets. We use ellipses (…) to indicate a continuation of the same pattern. A never-ending collection of elements is called an infinite set.

Example: Determine whether the sets are finite or infinite.
  a) $\{5, 8, 11, ..., 35\}$
  b) $\{1, 4, 9, 16, 25...\}$
  c). $\{x \mid x^2 = 121\}$

Solution:

a) The set begins with 5, ends at 35. We can also determine all the elements in between by adding 3 to the previous number. The set $\{5, 8, 11, ..., 35\}$ is then finite.

b) The set $\{0, 1, 4, 9, 16, 25, ...\}$ is the set of square integers. The list goes on and on as the ellipses indicate. Although the first element of the list is indicated, the list is never-ending and therefore we say that $\{0, 1, 4, 9, 16, 25, ...\}$ is an infinite set.

The solutions to the equation of $x^2 = 121$ are the members that belong to the set $\{x \mid x^2 = 121\}$. The only solutions to this equation are $x = 11$ or $x = -11$. Therefore $\{x \mid x^2 = 121\} = \{-11, 11\}$ which is a finite set.

EXERCISES 6.3

Identify each set as finite or infinite.

1. S={1,3,a,b,c,d}

2. T={1,2,3,...,100}

3. J={1,3,5,7,9,...}

4. A={x|x ∈ N and x ≤ 50}

5. A={x|x ∈ Z and x < 12}

6. P={x|x ∈ R and 2 ≤ x ≤ 5}

7. S={x|x ∈ Q and x ≥ $\frac{1}{2}$}

8. S={x| x is a time of day}

9. The set of all fish in Lake Ontario on July 1, 2004 at 7:00 pm.

10. B={x|x ∈ W and 2x+10<5}

Determine the cardinality of the finite sets below. If the set is infinite, say so and omit discussion on cardinality.

11. A={1,a,b,}

12. T={1,2,3,...,50}

13. J={x|x is an even natural number less than 12}

14. A={x| x is an odd whole number less than -5}

15. A={x|x ∈ N and x < 11}

16. P={x|x ∈ Z and 2 ≤ x ≤ 5}

17. B={x|x ∈ W and 2x+10<5}

18. The set of all whole numbers less than 0.

19. The set consisting of the real numbers that satisfy the equation x-8=15 .
20. The set of all states in the U.S.

Determine whether the given pairs of sets are equal, equivalent, or neither.

21. {a,b,c,d}, {b,d,a,c}

22. {1,2,3,4,5}, {6,7,8,9,10}

23. $\{x \mid x-7=5\}$, $\{x \mid x \in N \text{ and } x^2 = 144\}$

24. {5,5,10,10,15,15,...25}, {5,10,15,20,25}

25. {one, two, three, four,..., fifteen}, {1,2,3,4,...15}

Section 6.4: Set Operations

We already discussed one operation of sets earlier in the chapter when we introduced the notion of the complement. In this section, we concentrate on two more set operations called unions and intersections. Unions and intersections play an important role in set theory for many reasons we will encounter in this section. One of those reasons is directly related to creating, organizing, and retrieving data or tables of data. Organizing and maintaining large amounts of data is common in information technology. Databases are used to maintain all this data. The process of retrieving this data is where relational databases becomes useful when used in conjunction with the an appropriate human computer interface. After all, having data available to us is of little or no use unless the data returned is the data we request.

Telling the computer to give us the data we are interested in and how to present it to us is crucial in a society where time and efficiency translate into money. The data we are discussing here serve as the elements of a set(table). The table that contains the data is similar to a set that contains elements. Doing this search or query of data efficiently is something an employer will appreciate. One language of a human-computer interface to extract data is called SQL (Structured Query Language). You can use SQL to retrieve information you need from the tables. You may learn more about this in any database textbook.

Combining tables in a database is similar to uniting sets in set theory. We begin this section by discussing the union of sets.

Objective A: Union of Sets

Suppose you want to throw a reunion party for your high school and college friends. Assume you kept a record of all your high school friends separate from your college friends. You now need a combined list so you unite the two lists of names and create a new list that contains all the names of your friends. Also note that you would not list a friend twice that happened to attend both high school and college with you. The scenario above outlines the notion of the union of sets. The union of sets deals with combining sets with other sets to create new sets. The mathematical definition of "combining" lists or sets is given below.

> **Definition:** The **union** of two sets A, B, is denoted and defined by
> $A \cup B = \{x \mid x \in A \text{ or } x \in B\}$, i.e. $A \cup B$ is the set that contains all the elements belonging to set A or to set B.

Example 1: Find $A \cup B$ given that $A = \{a, b, c, d\}$ and $B = \{1, a, 2, c\}$

Solution: The union of these sets contains all the elements that are listed in either set A or set B without repetition. Therefore $A \cup B = \{1, 2, a, b, c, d\}$.

Important note: The union of these sets is itself a set and appropriate notation must be used.

Example 2: Given
$$A = \{1, 2, 4, 8, 16\}$$
$$B = \{0, 1, 2, 3, 4, 5, 6\}$$
$$C = \{-5, 0, 5, 10, 15\}$$
$$D = \{\ \}$$

Find: a) $A \cup B$

b) $(B \cup C) \cup A$

c) $A \cup D$

d). $B \cup B$

Solutions:

a) The set contains elements in either set without repetition. Therefore, $A \cup B = \{0, 1, 2, 3, 4, 5, 6, 8, 16\}$

b). First we find the set $(B \cup C)$, then take the union of this new set with set A. Therefore $B \cup C = \{-5, 0, 1, 2, 3, 4, 5, 6, 10, 15\}$ and $(B \cup C) \cup A = \{-5, 0, 1, 2, 3, 4, 5, 6, 8, 10, 15, 16\}$

c) The union of set A with the empty set D is given by $A \cup D = \{1, 2, 4, 8, 16\}$

d) The union of any set with itself results in the set itself. You can see this since $B \cup B = \{0, 1, 2, 3, 4, 5, 6\}$

Example 3: Find $\overline{(A \cup B)}$ given :
$$U = \{u, n, i, v, e, r, s\}$$
$$A = \{r, u, n\}$$
$$B = \{s, u, r, e\}$$

Solution: We need to first find the set $(A \cup B) = \{n, u, r, s, e\}$. Next, we need to find the complement of $(A \cup B)$, i.e. $\overline{(A \cup B)} = \{i, v\}$

We can illustrate the union of two sets, $A \cup B$ by the Venn diagram below.

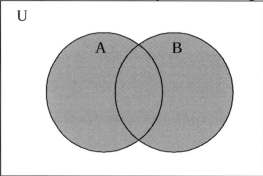

Note that the shaded region corresponds to all things that are common to either A or B (or both).

Objective B: Intersection of Sets

Suppose as in the earlier reunion example, you wanted to invite only the friends that attended both high school and college with you. You can go through the list of high school friends and see which names also occur on the list containing your college friends. The new list contains the names that were common to both your high school list and your college list. The intersection of sets deals precisely with this concept. The mathematical definition is given as:

Definition: The **intersection** of two sets A, B denoted and defined by
$A \cap B = \{x \mid x \in A \text{ and } x \in B\}$, i.e. $A \cap B$ is the set that contains all the elements that are in both set A and set B.

Example 1: Find $A \cap B$ given that $A = \{a, b, c, d\}$ and $B = \{1, a, 2, c\}$.

Solution: We are interested only in the elements that are common to both set A and B. The only elements that are common to both of these sets are a and c. Therefore $A \cap B = \{a, c\}$.

Example 2: Given
$$A = \{1,2,4,8,16\}$$
$$B = \{0,1,2,3,4,5,6\}$$
$$C = \{-5,0,5,10,15\}$$
$$D = \{\ \}$$

Find: a) $A \cap B$

b) $(B \cap C) \cap A$

c) $(A \cup B) \cap C$

d) $B \cap B$

Solution:

a). Set $A \cap B$ is the set that contains elements that set A and set B have in common. Therefore, $A \cap B = \{1,2,4\}$

b). We first find the set $(B \cap C)$, and find the intersection of this new set and set A. Therefore, $(B \cap C) \cap A = \{0,5\} \cap \{1,2,4,8,16\} = \{\ \}$

c). Here, we need to first find the union of A and B, then use this new set to finds the intersection with set C. Therefore,
$(A \cup B) \cap C = \{0,1,2,3,4,5,6,8,16\} \cap \{-5,0,5,10,15\} = \{0,5\}$

d). The intersection of any set with itself is the set itself. Hence,
$B \cap B = \{0,1,2,3,4,5,6\}$

Example 3: Find $\overline{A} \cap \overline{B}$ given
$$U = \{m,i,r,a,c,l,e\}$$
$$A = \{a,l,e\}$$
$$B = \{c,a,r,e\}$$

Solution: We first need to find $\overline{A}$, then $\overline{B}$ and then list the elements common to both sets. $\overline{A} = \{m,i,r,c\}$ and $\overline{B} = \{m,i,l\}$. $\overline{A} \cap \overline{B} = \{m,i\}$

We can illustrate the intersection, $A \cap B$, of any two sets, A, B by the Venn diagram below.

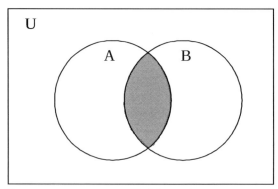

The shaded region of the diagram represents all things that are common to both A and B.

Example 3: Find $A \cap (B \cup C)$ given
$$A = \{1, 2, 3, 5\}$$
$$B = \{2, 4, 6, 8, 10, 12\}$$
$$C = \{1, 2, 3, 5, 8, 13, 21\}$$

Solution: We need to first find the union of B and C i.e.
$B \cup C = \{1, 2, 3, 4, 5, 6, 8, 10, 13, 12, 21\}$. $A \cap (B \cup C)$ is the collection of all elements that are common to set A and to set $B \cup C$ Therefore,
$A \cap (B \cup C) = \{1, 2, 3, 5\}$

De Morgan's Laws for Sets

Earlier in the text when we studied logic, we encountered De Morgan's Laws. Those laws established two logically equivalent statements that involved negation, conjunction, and disjunction. Similarly to De Morgan's Laws for logic, there are De Morgan's Laws that are used in set theory. In set theory, De Morgan's Laws involve the complement, union, and intersection. De Morgan's Laws as applied to sets is given below and a visual proof with Venn diagrams is given after a few examples.

De Morgan's Laws for Set Theory:

1. $\overline{A \cap B} = \overline{A} \cup \overline{B}$
2. $\overline{A \cup B} = \overline{A} \cap \overline{B}$

Example 4: Use De Morgan's Laws to show that $\overline{\overline{A \cap B}} = A \cup \overline{B}$

Solution: We will apply the first law and proceed as follows:
$$\overline{\overline{A \cap B}} = \overline{\overline{A} \cup \overline{B}} = A \cup \overline{B}$$

Notice that we made use of the double complement here.

Visual Proof Of De Morgan's Laws

We will use Venn diagrams to prove one of De Morgan's Laws. We will show that $\overline{A \cap B} = \overline{A} \cup \overline{B}$ and leave the proof of the other law as an exercise.
We begin by constructing a Venn diagram with two regions. We need to show the shading of $\overline{A \cap B}$ coincides with the shaded region of $\overline{A} \cup \overline{B}$.

To construct the Venn diagram for $\overline{A \cap B}$, we first shade the region of $A \cap B$, then then region outside this shading is $\overline{A \cap B}$.
Step 1: $A \cap B$

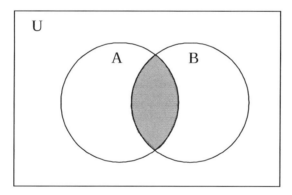

Step 2: $\overline{A \cap B}$

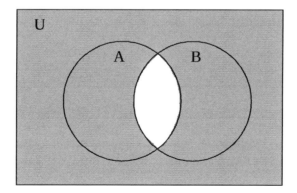

352

To show the equivalence in De Morgan's Laws, we now shade the region of $\overline{A} \cup \overline{B}$.

Step 1: $\overline{A}$

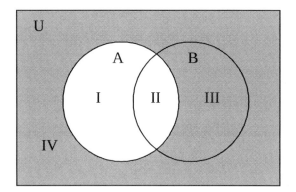

Step 2: $\overline{B}$

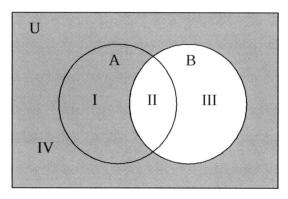

The union of these two sets is contains the shaded region in either $\overline{A}$ or $\overline{B}$. You can see the region shaded for $\overline{A}$ is III and IV, and the region shaded for $\overline{B}$ is I and IV. The shaded region of the union therefore consists of regions I, III, and IV. The diagram is given below.

Step 3: $\overline{A} \cup \overline{B}$

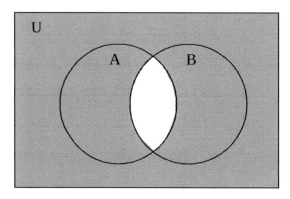

Notice that the shaded region for $\overline{A \cap B}$ and $\overline{A} \cup \overline{B}$ are the same. Therefore we can see that $\overline{A \cap B} = \overline{A} \cup \overline{B}$

Objective C: Difference of Sets:

Definition: The **difference** of two sets A, B, is denoted and defined by $A - B = \{x \mid x \in A \text{ and } x \notin B\}$, i.e. $A - B$ is the set that contains all the elements of set A that are not in set B.

Example 1: Find $A - B$ given

$$A = \{a, b, c, d\}$$
$$B = \{1, a, 2, g\}$$

Solution: We need to find the **difference** of two sets. We are interested in all the elements of A that are not in B. Therefore $A - B = \{b, c, d\}$.

Note: proper set notation must be used since the difference of the two sets is also a set.

Example 2: Find $C - D$ given $C = \{1, 2, 4, 7, 8\}$ and $D = \{2, 3, 5, 7, 9, 11\}$

Solution: The elements in the set $C - D$ are the elements in C that are not in D. Therefore, $C - D = \{1, 4, 8\}$

Example 3: Find $D - C$ given $C = \{1, 2, 4, 7, 8\}$ and $D = \{2, 3, 5, 7, 9, 11\}$

Solution: The elements in the set $D - C$ are the elements in D that are not in A. Therefore, $D - C = \{3, 5, 9, 11\}$

Example 4: Find $W - N$ where W and N denote the set of whole numbers and set of natural numbers respectively.

Solution: The set contains all the elements of W that are not in N. The only element in W that is not in N is 0. i.e. $W - N = \{0\}$

Venn Diagram Illustration of the Difference of Two Sets

The following Venn diagram outlines the region of the set $A - B$

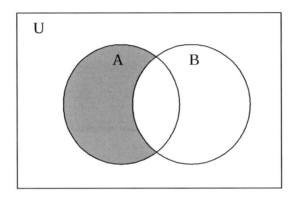

Objective D: Applications of Set Cardinality

Many types of problems require gathering and analyzing data. In this section, we will study how to analyze surveys. For example, a survey on the type of music a college student listens to might benefit advertisers and radio stations in the selection of their program guide. Suppose a survey of 500 students is conducted, and the survey finds that 230 students listen to techno music, 150 students listen to hip-hop music, and 75 listen to both techno and hip-hop. One question might be: how many students do not listen to either? While another question might be: how many students listen only to techno. You may be inclined to incorrectly answer 230 for the latter question. Problems such as this can be analyzed using Venn diagrams and the cardinality of sets. We will use Venn diagrams to analyze this type of data.

Example 1: A survey of 500 students for their preference of music finds that:
   230 listen to techno
   150 listen to hip-hop
   75 listen to both techno and hip-hop
   a) How many students listen to techno only?
   b) How many student listen to hip-hop only?
   c) How many students listen to neither?

Solution: We will construct a Venn diagram with the cardinality of the universal set depicting the total number of students surveyed. We will also include two subsets, depicting the cardinality of the set of students that listen to techno, and hip-hop.

If we let U denote the universal set of all possible responses of the survey, the set H to denote hip-hop, and the set T to denote techno, then the given information is:

n(U) = 500
n(T) = 230
n(H) = 150
n(T∩H) = 75

Translating this information onto the Venn diagram yields:

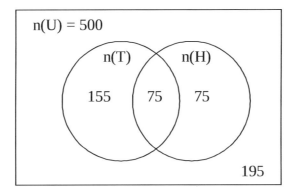

The region entirely surrounded as techno has 230 students, and the region entirely surrounded by hip-hop has 150 students. It was important to separate the region to indicate the quantity of students that listen to both. The region outside the circular region represents the number of students surveyed that do listen to neither techno nor hip-hop. The number is found by taking the total surveyed and subtracting the sum of all the regions. That is, $500 - (155 + 75 + 75) = 195$.

We are now ready to answer the questions.
    a) 155 students listen to techno only.
    b) 75 students listen to hip-hop only.
    c) 195 students listen to neither.

An important starting point is where most of the information is given. In this case, the intersection of both regions is a nice place to start. Generally speaking, if the cardinality of the intersecting region is given, we will begin there and work outwards taking into account how many of the members are already accounted for and determining the remainder. For example, we determined that 155 students listen to only techno since 75 of them were already accounted for in the intersecting region. This intersecting region is also part of the techno region.

Some surveys may request information on more than two possibilities. For example, suppose in the survey above, listening to rock and roll was another choice students could make. It is possible that there may be students that listen to all three types of music. The Venn diagram then must include a region that reflects the interesting of what would be three subsets. The general Venn diagram with three subsets is given as:

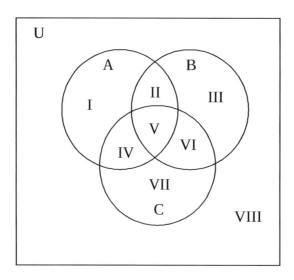

The following sets and the corresponding regions are given in the table below.

Set	Regions
A	I, II, IV, V
B	II, III, V, VI
C	IV, V, VI, VII
$A \cap B \cap C$	V
$A \cap B$	II, V
$A \cap C$	IV, V
$B \cap C$	V, VI
$A \cup B$	I, II, III, IV, V, VI
$A \cup C$	I, II, IV, V, VI, VII
$B \cup C$	II, III, IV, V, VI, VII
$A \cup B \cup C$	I, II, III, IV, V, VI, VII
$\overline{A \cup B \cup C}$	VIII

Example 2: Suppose 300 people were surveyed on the ice-cream flavors they like to order on a cone. The survey reported that:
120 liked vanilla
180 preferred chocolate
90 preferred strawberry
27 preferred vanilla and chocolate
30 preferred vanilla and strawberry
55 preferred strawberry and chocolate
15 preferred all three

a) How many students did not prefer any of these flavors?
b) How many students preferred only chocolate?
c) How many students preferred vanilla but not strawberry.

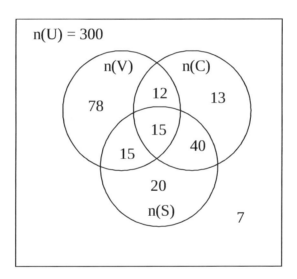

Solution: For simplicity we begin by assigning the set V to vanilla, C to denote chocolate, and S to denote strawberry. The information gioven above can be summarized as follows:

$$n(U) = 300$$
$$n(V) = 120$$
$$n(C) = 180$$
$$n(S) = 90$$
$$n(V \cap C) = 27$$
$$n(V \cap S) = 30$$
$$n(S \cap C) = 55$$
$$n(C \cap V \cap S) = 15$$

Using the table above for guidance of the region and beginning in region 5, we get:

a) This quantity is found in region 8, $\overline{V \cup C \cup S}$. Since $n(\overline{V \cup C \cup S}) = 7$,, only 7 students did not prefer any of these flavors.

b) The students that prefer only chocolate implies chocolate and nothing else. This corresponds to region 3, so there are 113 students that prefer only chocolate.

c) The region for students that prefer vanilla but not strawberry corresponds to regions 1,2. the sum of these regions is 90. Therefore 90 students prefer vanilla but not strawberry.

EXERCISES 6.4

Perform the indicated operations if:

$U = \{a,b,c,d,e,f,g,h,i\}$

$A = \{a,c,d,g\}$

$B = \{e,h,i\}$

$C = \{a,b,d,g\}$

1. $A \cup B$
2. $\overline{A} \cap B$
3. $A \cup (B \cap C)$
4. $B \cap \overline{(A \cup C)}$
5. $C - U$
6. $U - (A \cup C)$
7. $\overline{B} \cup \overline{A}$
8. $\overline{\overline{B} \cap \overline{A}}$
9. $C \cup (B \cap \overline{U})$
10. $(A \cap B) \cup (B - C) \cup \overline{(A - B)}$

Shade the region of the Venn diagram corresponding to the given sets.

11. $\overline{A \cup B}$

12. $\overline{B} \cup \overline{A}$

13. $(A \cup B) \cap C$

14. $\overline{(A \cap \overline{B})} \cup C$

15. $(A - B) \cap C$

16. $\overline{\emptyset} \cap (A \cup B \cup C)$

17. $U - (A \cup B \cup C)$

18. $U - (A \cap B \cap C)$

19. $\overline{A \cup B}$

20. $\overline{A} \cap \overline{B}$

Use Venn diagrams to answer the following survey problems.

21. Pizza Parlor: 75 people went to the pizza parlor and ordered a pizza pie. 30 ordered sausage on their pizza, 40 ordered pepperoni on their pizza, and 10 ordered a pepperoni and sausage on their pizza. How many people did not order either sausage or pepperoni

22. Politics: A survey of 675 people by a beverage marketing corporation find that 225 people drink carbonated soda, 175 people drink non carbonated soda, and 45 people do not drink any type of soda. How may people drink both types?

23. Courses: A survey of students on whether they are taking a mathematics course, a biology course, and an English course. We find that 20 students take English, 7 students take English and Biology, 8 students take Math and English, 6 students take Biology only, 2 students take Math only, 7 students take neither of the courses, and 5 students take all three. How many students take Math and Biology but not English?

24. Computers: In a random survey of 230 people, it was found that 102 people owned a desktop computer, 95 owned a laptop computer, 57 owned a tablet PC, 20 owned a notebook and a desktop computer, and 25 owned a notebook and a tablet PC, 12 owned a desktop PC and a tablet PC, and 5 owned all three. How many people do not own any of these three computers?

Appendix A

BASIC Statements

The statements that you see in the following pages are the ones that we use the most in this course. There are many more statements that may be given to you during the semester by your instructor.

Functions

ABS
returns the absolute value of a number.

Example: **PRINT ABS (25-65)**   Output is: 40

ASC
returns the ASCII code for the first character in a string expression.

Example: **PRINT ASC("A")**   Output is: 65

CHR
returns the character corresponding to a specified ASCII code.
If *n* is a number from 0 to 255, then **CHR$(*n*)** is the character in the ASCII table associated with *n*.

Example: **PRINT CHR$(97)**   Output is: a (lowercase a)

INT
returns the largest integer less than or equal to a numeric expression

Example: **PRINT INT(12.54), INT(-23.2)**   Output is: 12   -24

LEN
returns the number of characters in a string or the number of bytes required to store a variable.

Example: **PRINT LEN("A QBasic Statement")**   Output is: 18

RND
returns a single precision random number between 0 and 1.

SGN
returns a value indicating the sign of a numeric expression
(1 if the expression is positive, 0 if it is zero, or -1 if it is negative).

Example: **PRINT SGN(1), SGN(-25), SGN(0)**   Output is: 1   -1   0

SQR
returns the square root of a numeric expression.

Example: **PRINT SQR(25)**   Output is: 5

STR$
returns a string representation of a number.

TAB
The function **TAB(*n*)** is used in PRINT statements to move the cursor to position *n* and placed blank spaces in all skipped-over positions. If *n* is less than the cursor position, then the cursor is moved to the *n*th position of the next line.

MOD
(Operator)

divides one number by another and returns the remainder.
   *numeric-expression*1 **MOD** *numeric-expression*2   where
   *numeric-expression*1 and *numeric-expression*2 are numeric expressions.
Real numbers are rounded to integers.

Example: **PRINT 19 MOD 3.7**   Result: 3.7 is rounded to 4
                                        Output is: 3

More BASIC Statements

AND
(Logical Operator)

The logical expression *condition*1 AND *condition*2 is true only if both *condition*1 and *condition*2 are true.

Example: **(1<5) AND ("edf" > "e")** is true since both conditions are true.
("apple">"ape") AND ("dog"<"cat") is false since "dog"<"cat" is false.

CALL
(Statement)

A statement of the form **CALL *SubprogramName*(argumentlist)** is used to execute the named subprogram, passing to variables and values in the list of arguments. Separate multiple arguments with commas. Specify array arguments with the array name followed by empty parenthesis. The value of a variable argument may be changed by the subprogram unless the variable is surrounded by parenthesis. After all the statements in the subprogram are executed, program execution resumes with the statement following **CALL**.

CLS
(Statement)

Clears the screen and positions the cursor at the upper left of the screen.

DATA
(Statement)

specifies values to be read by **READ** statement. One or more numeric or string constants specifying the data to be read, if more, commas are used to separate them. String constants containing commas, colons, or leading or trailing spaces are enclosed in quotation marks(" ").

Example: DATA   *constant,constant,....*

READ
(Statement)

Reads the values from **DATA** and assigns them to variables.

Example: READ *grade1,grade2,grade3*   READ   *date$, temperature*
              DATA   95,60,83   DATA   "January 5, 2000", 43

              READ *name$,* salary
              DATA   Grover, 200

RESTORE
(Statement)
allows **READ** to reread values in specified **DATA** statements.

DIM
(Statement)
declares an array or specifies a data type for a nonarray variable.

The statement **DIM** *arrayName(lower* **to** *upper)* declares an array with subscripts ranging from *lower* to *upper*, inclusive, where *lower* and *upper* are in the normal integer range of -32,768 to 32,767.

A statement **DIM** *arrayName(lower1* **to** *upper1, lower2* **to** *upper2)* declares a doubly subscripted, or two-dimensional, array.

A statement **DIM** *arrayName(upper)* assumes the lower bound to be zero.

Statement **DIM** *variableName* **AS** *variableType* where *variableType* can be INTEGER, LONG, SINGLE, DOUBLE, STRING, or a user-defined data type).

Some data types:
INTEGER   requires two bytes of memory and can hold whole numbers from -32,768 to 32,767.
LONG   requires four bytes of memory and can hold whole numbers from -2,147,483,648 to 2,147,483,647.
SINGLE   requires four bytes of memory and can hold 0, the numbers from 1.40129×10^{-45} to 3.40283×10^{38} with at most seven significant digits, and the negative of these numbers.
DOUBLE   requires eight bytes of memory and can hold 0, the numbers from 4.94065×10^{-324} to $1.797693134862316 \times 10^{308}$ with at most 17 significant digits, and the negative of these numbers.

Examplse: **DIM** grade(1 to 10)   declares an array where 10 grades will be placed

       DIM A(1 to 3, 1 to 5)   declares a 3 by 5 two dimensional array or matrix.
       DIM num AS INTEGER   declares the variable **num** as integer.

DO/LOOP *see repetitions*

END
(Statement)

end a program, procedure, block, or user-defined data type.

INPUT
(Statement)

The statement **INPUT** *var* causes the computer to display question mark and to pause until the user enters a response. This response is then assigned to the variable *var*.

A statement of the form **INPUT "***prompt***";** *var* inserts a prompting message before the question mark.

A statement of the form **INPUT "***prompt***",** *var* displays the message without the question mark.

In each of these statements *var* may be replaced by a number of variables separated by commas. After the user enters the values requested, each value is assigned to the corresponding variable.

Example: **INPUT "Enter your name"; name$**   Output is : Enter your name?
                                                                         [User then enters the name]

LET
(Statement)

assigns the value of an expression to a variable.
    LET *variable = expression* (The word **LET** may be omitted)

Variable names can consist of up to 40 characters and must begin with a letter. Valid characters are A-Z, 0-9, and period(.).

Example: **LET** *num5* = (3+5)^2   Result: The content of *num5* will be 64.

PRINT
(Statement)

writes data to the screen or to a file.

PRINT *expressionlist* (separated by comma or semi-colon)

expressionlist is a list of one or more numeric or string expressions

comma or semicolon determines where the next output begins:

, means print at the start of the next print zone

(Each print zone is 14 characters wide.)

; means print immediately after the last value.

Example: **PRINT "This is a test "; "only a test"** (separated by a semi-colon)

Output: This is a test only a test

PRINT "This is a test", "only a test" (separated by a comma)

Output: This is a test   only a test

PRINT USING
(Statement)

Writes formatted output to the screen or to a file.

PRINT USING *formatstring$; expressionlist* (with , or ;)

Example: **PRINT USING a$, n**

Symbol	Meaning	n	a$	Result
#	Digit position	1234.56	"######"	1235
		765	"######"	765
		783.4	"######"	783
		12873	"####"	%12873 (overflow)
.	Decimal placement	345.2	"###.#"	345.2
,	Print a comma before every third digit to the left of decimal	1234567	"#######,"	1,234,567
+	Print the sign of a number before or after the number	123	"+####"	+123
		-123	"+####"	-123
		123	"####+"	123+
		-123	"####+"	123-
-	Print the sign of a number before or after negative numbers only	123	"-####"	123
		-123	"-####"	-123
		123	"####-"	123

Symbol	Meaning	n	a$	Result
		-123	"####-"	123-
$	Print a $ sign as the first character of the field	45.25	"$####.##"	$ 45.25
$$	Print a $ sign immediately before the first digit displayed	45.25	"$$####.##"	$45.25
**	Fill print field with asterisks before number	45.25	"**####.##"	****45.25
$**	Fill print field with a dollar sign followed by asterisks before number	45.25	"$**###.##"	$***45.25
**$	Fill print field with asterisks followed by a dollar sign before number	45.25	"**$###.##"	***$45.25
!	Print only the first character of the string		"Suffolk"	S
\ \	Print first n characters, where n is the number of blanks between the two slashes + 2		"Suffolk" \ \ (three spaces between slashes)	Suffo
&	Print the entire string		"Suffolk"	Suffolk

REM or '
remark, this allows comments to be inserted in a program (not executed)

SWAP
(statement)

Interchanges the values of memory locations.

SWAP *variable1*, *variable2*   Interchanges the values of *variable1* and *variable2*.

Example: LET   A = 4
      LET   B = 5
   Swap A, B   OUTPUT: A = 5 and B = 4

REPETITION STRUCTURE

1. Pre-test Loop

DO WHILE *condition* ....
    ...
    ... } Statements or actions that are being repeated
    ...
LOOP

This loop structure does the testing at the beginning of the loop, if the condition is met, then the statements within the loop will be executed. The repetition will stop once the condition is no longer true.

2. Post-test Loop

DO ...
    ... } Statements or actions that are being repeated
    ...
LOOP UNTIL *condition*

This loop structure does the testing at the end of the loop, if the condition is satisfied, then the loop is terminated, otherwise, loop continues.

3. Loop is executed a preset number of times

FOR variable = initial **TO** limit **STEP** increment
    ...
    ... } Statements or actions that are being repeated
    ...
NEXT variable

In this structure, **variable** serves as the counter. It is initialized to the number specified by **initial**, incremented by the number specified after the word **STEP**, and ends the loop when **variable** is equal to or more than what is specified by **limit.** If **STEP** is not specified, the default is the counter will be incremented by 1. **STEP** could also decrement instead of increment.

There are other forms of loops, however, we are going to use the above three forms.

DECISION STRUCTURE

1. **IF** **(**single line**)**

IF *condition* **THEN**

A statement of the above form causes the program to take the specific action if *condition* is true. Otherwise, execution continues at the next line.

2. **IF (**single line**)**

IF *condition* **THEN** *action1* **ELSE** *action2*

A statement of the above form causes the program to take *action1* if *condition* is true and *action2* if *condition* is false.

3. **IF (**block**)**

IF *condition1* **THEN**
Statement1
IF *condition2* **THEN**
Statement2
IF *condition3* **THEN**
. . . . .
. . . . .
END IF

A block of statements such as the one above indicates that the group of statements between **IF** and **END IF** are to be executed only when *condition* is true.

If the group of statements is separated into two parts by an **ELSE** statement, then the first part will be executed if the *condition* is true and the second part will be executed when the *condition* is false.

4. SELECT CASE

SELECT CASE *testexpression*
    CASE *expressionlist1*
       Statementblock1
    CASE *expressionlist2*
       Statementblock2

    CASE *expressionlist3*
       Statementblock3
    . . . . .
    . . . . .
END SELECT

The **SELECT CASE** statement provides a compact method of selecting for execution one of the several blocks of statements based on the value of an expression.

Testexpression can be any numeric or string expression. *Expressionlist1*, *Expressionlist2*,.... are one or more expressions to match the *testexpression*. *Statementblock1*, *Statementblock2*,. . . . are one or more statements on one or more lines which are to be executed if the *expressionlist* is the case.

Appendix B

Geometric Formulas

Lengths

The unit of measurement for a length is always a single unit measure such as inches (in), feet (ft), centimeters (cm), meter (m), miles (mi), kilometers (Km) etc.

<u>Perimeter of a rectangle:</u>

$$p = 2 \times L + 2 \times W$$

<u>Circumference of a Circle:</u>
$$c = 2 \times \pi \times r$$
$$ = \pi \times d$$

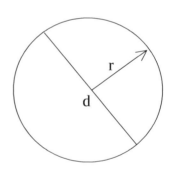

<u>Length of the hypotenuse of a right triangle:</u>
$$c = \sqrt{a^2 + b^2} \text{ or } c^2 = a^2 + b^2$$

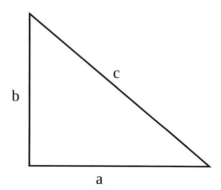

Areas

The unit of measurement for area is always a length squared, since area is obtained by multiplying two lengths together. e.g. inches squared (in²), feet squared (ft²), centimeters squared (cm²), meters squared (m²), square miles (mi²), square kilometers (Km²), etc.

Rectangle: $A_R = L \times W$

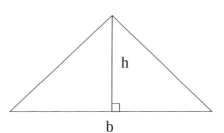

Triangle: $A_T = \dfrac{1}{2} \times b \times h$

Circle: $A_C = \pi \times r^2$

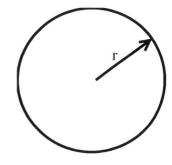

Semicircle: $A_{SC} = \dfrac{1}{2} \times \pi \times r^2$

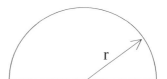

Surface area of a sphere:

$A_{Sphere} = 4 \times \pi \times r^2$

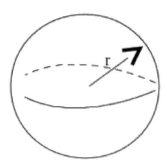

Surface area of a rectangular solid:

$A_{RectSolid} = 2(L \times H) + 2(L \times W) + 2(H \times W)$

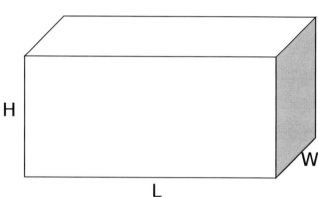

Volumes

The unit of measurement for volume is always a length cubed, since volume is obtained by multiplying three lengths together. e.g. inches cubed (in^3), feet cubed (ft^3), centimeters cubed (cm^3), meters cubed (m^3), cubic miles (mi^3), cubic kilometers (Km^3), etc.

Rectangular solid:
$V_{Rect} = L \times W \times H$

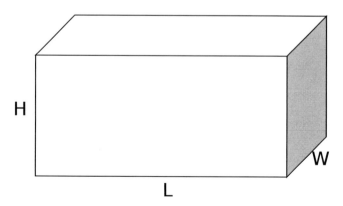

Sphere:
$$V_s = \frac{4}{3} \times \pi \times r^3$$

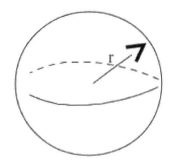

Cylinder:
$$V_{Cyl} = \pi \times r^2 \times h$$

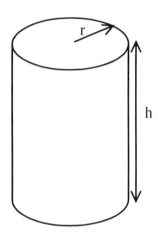

Right circular cone:
$$V_{Cone} = \frac{1}{3} \times \pi \times r^2 \times h$$

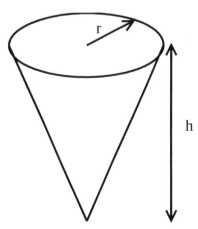

Appendix C

ASCII Table

Dec	Oct	Hex	Binary	Char	Description
000	000	00	00000000	^@	NULL- Null
001	001	01	00000001	^A	SOH - Start of Header
002	002	02	00000010	^B	STX - Start of text
003	003	03	00000011	^C	ETX - End of Text
004	004	04	00000100	^D	EOT - End of Transmission
005	005	05	00000101	^E	ENQ - Enquiry
006	006	06	00000110	^F	ACK - Acknowledge
007	007	07	00000111	^G	BEL - Bell
008	010	08	00001000	^H	BS - Backspace
009	011	09	00001001	^I	HT - Horizontal Tab
010	012	0A	00001010	^J	LF - Line Feed
011	013	0B	00001011	^K	VT - Vertical Tab
012	014	0C	00001100	^L	FF - Form Feed
013	015	0D	00001101	^M	CR - Carriage Return
014	016	0E	00001110	^N	SO - Shift Out
015	017	0F	00001111	^O	SI - Shift In
016	020	10	00010000	^P	DLE - Data Link Escape
017	021	11	00010001	^Q	DC1 - XON
018	022	12	00010010	^R	DC2 - Not used
019	023	13	00010011	^S	DC3 - XOFF
020	024	14	00010100	^T	DC4 - Not used
021	025	15	00010101	^U	NAK - Negative acknowledge
022	026	16	00010110	^V	SYN - Synchronous Idle
023	027	17	00010111	^W	ETB - End of Transmission Block
024	030	18	00011000	^X	CAN - Cancel
025	031	19	00011001	^Y	EM - End of Medium
026	032	1A	00011010	^Z	SUB - Substitute
027	033	1B	00011011	^[	ESC - Escape
028	034	1C	00011100	^\	FS - File Separator
029	035	1D	00011101	^]	GS - Group Separator
030	036	1E	00011110	^^	RS - Record Separator
031	037	1F	00011111	^_	DEL - Delete
032	040	20	00100000	SPC	Space
033	041	21	00100001	!	Exclamation mark
034	042	22	00100010	"	Quotes

Dec	Oct	Hex	Binary	Char	Description
035	043	23	00100011	#	Hash
036	044	24	00100100	$	Dollar
037	045	25	00100101	%	Percent
038	046	26	00100110	&	Ampersand
039	047	27	00100111	'	Apostrophe
040	050	28	00101000	(	Open bracket (Parenthesis)
041	051	29	00101001	)	Close bracket (Parenthesis)
042	052	2A	00101010	*	Asterisk
043	053	2B	00101011	+	Plus
044	054	2C	00101100	,	Comma
045	055	2D	00101101	-	Dash
046	056	2E	00101110	.	Full stop
047	057	2F	00101111	/	Slash
048	060	30	00110000	0	Zero
049	061	31	00110001	1	One
050	062	32	00110010	2	Two
051	063	33	00110011	3	Three
052	064	34	00110100	4	Four
053	065	35	00110101	5	Five
054	066	36	00110110	6	Six
055	067	37	00110111	7	Seven
056	070	38	00111000	8	Eight
057	071	39	00111001	9	Nine
058	072	3A	00111010	:	Colon
059	073	3B	00111011	;	Semi-colon
060	074	3C	00111100	<	Less-than
061	075	3D	00111101	=	Equals
062	076	3E	00111110	>	Greater than
063	077	3F	00111111	?	Question mark
064	100	40	01000000	@	At
065	101	41	01000001	A	
066	102	42	01000010	B	
067	103	43	01000011	C	
068	104	44	01000100	D	
069	105	45	01000101	E	
070	106	46	01000110	F	
071	107	47	01000111	G	
072	110	48	01001000	H	
073	111	49	01001001	I	
074	112	4A	01001010	J	
075	113	4B	01001011	K	
076	114	4C	01001100	L	
077	115	4D	01001101	M	
078	116	4E	01001110	N	

Dec	Oct	Hex	Binary	Char	Description
079	117	4F	01001111	O	
080	120	50	01010000	P	
081	121	51	01010001	Q	
082	122	52	01010010	R	
083	123	53	01010011	S	
084	124	54	01010100	T	
085	125	55	01010101	U	
086	126	56	01010110	V	
087	127	57	01010111	W	
088	130	58	01011000	X	
089	131	59	01011001	Y	
090	132	5A	01011010	Z	
091	133	5B	01011011	[	Open Square bracket
092	134	5C	01011100	\	Backslash
093	135	5D	01011101	]	Close Square bracket
094	136	5E	01011110	^	Caret/hat
095	137	5F	01011111	_	Underscore
096	140	60	01100000	`	Accent grave
097	141	61	01100001	a	alpha
098	142	62	01100010	b	bravo
099	143	63	01100011	c	charlie
100	144	64	01100100	d	delta
101	145	65	01100101	e	echo
102	146	66	01100110	f	foxtrot
103	147	67	01100111	g	golf
104	150	68	01101000	h	hotel
105	151	69	01101001	i	india
106	152	6A	01101010	j	juliett
107	153	6B	01101011	k	kilo
108	154	6C	01101100	l	lima
109	155	6D	01101101	m	mike
110	156	6E	01101110	n	november
111	157	6F	01101111	o	oscar
112	160	70	01110000	p	papa
113	161	71	01110001	q	quebec
114	162	72	01110010	r	romeo
115	163	73	01110011	s	sierra
116	164	74	01110100	t	tango
117	165	75	01110101	u	uniform
118	166	76	01110110	v	victor
119	167	77	01110111	w	whiskey
120	170	78	01111000	x	x-ray
121	171	79	01111001	y	yankee

Dec	Oct	Hex	Binary	Char	Description
122	172	7A	01111010	z	zulu
123	173	7B	01111011	{	Open Brace
124	174	7C	01111100	|	Pipe
125	175	7D	01111101	}	Close Brace
126	176	7E	01111110	~	Tilde
127	177	7F	01111111	^?	DEL - Delete

Appendix D

Answers to Select Exercises

Chapter 1
1.2
1. a. {52, 496}
   b. {52, 0, 496}
   c. {-5, 52, 0, 496}
   d. {3.4, –5, 52, 0, 3/8, -.632, 496, -3.25}
   e. { $\sqrt{7}$,-Π}
2. a. {92, $\sqrt{25}$ =5}
   b. {92, $\sqrt{25}$ =5}
   c. {-6, 92, $\sqrt{25}$ }
   d. {-6, 1.5, ½, $\sqrt{25}$, -.62, 92, -8..33}
   e. { $\sqrt{3}$ }
3. a. {23, 567}
   b. {23, 567, 0}
   c. {23, -1. 567, 0}
   d. {23, -1, 1.2, 3/7, -5/6, 567, 0 –1.34}
   e. { $\sqrt{11}$, Π}
4. output: The sum of a and b is 15
5. output: The product of a and b is 35
6. output: The difference of a and b is 11

question	statement	meaning	truth value
7	3<3	three is less than three	false
8	4≤8	four is less than or equal to eight	true
9	-1>-4	negative one is greater than negative four	true
10	9=8	nine is equal to eight	false
11	10≠10	ten is not equal to ten	false
12	8≥-10	eight is greater than or equal to negative ten	true
13	5≥5	five is greater than or equal to five	true
14	-1≤-4	negative one is less than or equal to negative four	false
15	-2 > -12	negative two is greater than negative twelve	true
16	-6 < -6	negative six is less than negative six	false
17	-8 < -2	negative eight is less than negative two	true
18	7≠9	seven is not equal to nine	true

1.3

1. −5
2. 7
3. −3/4
4. 1/5
5. −2.9
6. 3.6
7. 5
8. 9
9. 1.23
10. 56
11. 7
12. 4.7
13. 3/8
14. 1/3
15. 0.45
16. 8.4
17. 8
18. −10
19. 6
20. 5
21. −17
22. −15
23. 2/5
24. 1/7
25. 1/24
26. 3/20
27. −2.8
28. −5.9
29. 2.9
30. 0
31. −21.7
32. 3
33. −4
34. −7
35. −4
36. −13
37. 13
38. −2
39. 17
40. −5/7
41. −1/5
42. 4/9
43. 31/36
44. −6.8
45. −1.8
46. 5.2
47. −24
48. 10
49. −12.9
50. −3/2
51. 9.86
52. −20.5
53. −1/4
54. 2
55. −5
56. 3
57. −4
58. 5/6
59. −1/2
60. −10
61. 16
62. 25
63. 1/36
64. −8
65. 16/49
66. 15/19
67. 1
68. 729
69. −1024
70. 194,481
71. 16
72. −125
73. 1/49
74. 27/8
75. 1
76. 1/81
77. 25
78. −1/31
79. −1/64
80. 25

1.4

1. 6
2. 11
3. 12
4. 10
5. −7
6. 2/3
7. 21/5
8. 0.2
9. −14
10. 15/17
11. 4
12. −5
13. 4.47
14. 8.54
15. −6.71
16. 0.89
17. −1.41
18. 10
19. 12
20. 32
21. 6
22. 5.477226
23. 5.848035

24. Associative law of addition
25. Distributive law of multiplication over addition
26. Commutative law of multiplication
27. Commutative law of addition
28. Identity law of addition
29. Commutative law of addition
30. Associative law of multiplication
31. Associative law of addition
32. Inverse law of multiplication
33. Distributive law of multiplication over addition
34. Amount = 112.5
35. Percent = 0.375 or 37.5%

36. Base = 205
37. Amount =84
38. Interest = 10   This means Peter will have to pay $200+$10 or $210 at the end of the year
39. 0.0625 or 6.25%
40. −5
41. 7
42. −13
43. 10
44. 167
45. 48
46. 159
47. −4
48. −5
49. 70
50. 8
51. 26
52. −25
53. 37
54. −7

1.5
1. .27
2. 4.24
3. 5.39
4. −10.44
5. (a) $5.28045 * 10^9$
   (b) $.528045 * 10^{10}$
6. (a) $9.67 * 10^{-5}$
   (b) $.967 * 10^{-4}$
7. (a) $1.0923 * 10^{11}$
   (b) $.10923 * 10^{12}$
8. (a) $5.5 * 10^4$
   (b) $.55 * 10^5$
9. (a) $5.34 * 10^{-8}$
   (b) $.534 * 10^{-7}$
10. (a) $1 * 10^{11}$
    (b) $.1 * 10^{12}$

11. 10,000,700
12. 49,900
13. .00005536
14. .0002005
15. 1,023,000
16. 77,440,000

1.6
1. 1
2. 3
3. 4
4. 2
5. 6
6. 3
7. 4
8. 0
9. 3
10. 1
11. 2
12. 4
13. -4
14. 0
15. 12
16. 23

Chapter 2
2.1
1. 2^{16} bytes
2. 2^{18} bytes
3. 56,000,000 bytes; 0.056GB
4. 1,024,000,000 bytes; 1.024 GB
5. Approximately 4.5 MiB; 4,610.1 KiB
6. 11 MiB; 11,269 KiB
7. 33,000,000,000 bytes; 33,000,000 KB; 33,000MB; 32,226,579KiB; 31,485MiB
8. 80,000,000,000 bytes; 80,000,000 KB; 80,000 MB; 78,125,040 KiB; 76,320MiB
9. 450,000,000 picoseconds; 450,000 nanoseconds; 450 microseconds
10. 0.895 nanoseconds; 0.000895 microseconds

11. 3×10^{11} picoseconds; 3×10^{8} nanoseconds
12. 14,750 picoseconds; 0.01475 microseconds
13. 24,000,000 picoseconds; 24,000 nanoseconds; 24 microseconds
14. 3,121,000 picoseconds; 3.121 microseconds
15. 4.6×10^{12} picoseconds; 4.6×10^{9} nanoseconds
16. 0.12438 nanoseconds; 0.00012438 microseconds
17. Approximately 0.3355 seconds
18. 0.1074 seconds
19. 360,000 bytes/sec
20. 3,600,000 bytes/sec
21. Approximately 83.3 seconds
22. 110.3 seconds
23. 18.75 seconds
24. 45.95 seconds
25. 96,000KiB
26. 54,000 KiB
27. 3 GB
28. 8 GB
29. 22 hours/failure
30. Approximately 12 hours/failure

2.2

1. | - | 9 | 4 | 5 | 9 | 3 | 0 | + | 1 | 6 |

2. | + | 1 | 1 | 7 | 8 | 5 | 2 | - | 0 | 6 |

3. | + | 1 | 2 | 3 | 5 | 6 | 7 | + | 0 | 7 |

4. | - | 6 | 2 | 9 | 4 | 4 | 0 | - | 0 | 6 |

5. | + | 6 | 4 | 5 | 3 | + | 1 | 2 |

6. | - | 1 | 2 | 1 | 3 | - | 0 | 9 |

7. | - | 5 | 3 | 9 | 6 | + | 0 | 4 |

8. | + | 3 | 8 | 3 | 2 | - | 0 | 9 |

9. Answers may vary

	Rounded to one digit	Rounded to two digits	Rounded to three digits	Rounded to four digits
10.	800	790	789	789.1
11.	-40	-35	-35.2	-35.23
12.	10	14	13.8	13.78
13.	-1,000	-1,000	-1,020	-1024
14.	$0.1*10^{-3}$	$0.12*10^{-3}$	$0.124*10^{-3}$	$0.1240*10^{-3}$
15.	$-1*10^{7}$	$-0.99*10^{7}$	$-0.988*10^{7}$	$-0.9875*10^{7}$
16.	$0.2*10^{6}$	$0.24*10^{6}$	$0.236*10^{6}$	$0.2356*10^{6}$
17.	$0.2*10^{-11}$	$0.25*10^{-11}$	$0.249*10^{-11}$	$0.2493*10^{-11}$

	Truncated at the decimal place	Truncated at one decimal place	Truncated at two decimal place	Truncated at three decimal place
18.	789	789.1	789.12	789.124
19.	-35	-35.2	-35.23	-35.238
20.	13	13.7	13.77	13.775
21.	-1,024	-1,024.0	-1,024.09	-1024.092

	Truncated at the decimal place	Truncated at one decimal place	Truncated at two decimal place	Truncated at three decimal place
22.	$0.1*10^{-3}$	$0.12*10^{-3}$	$0.123*10^{-3}$	$0.1239*10^{-3}$
23.	$-.9*10^{7}$	$-0.98*10^{7}$	$-0.987*10^{7}$	$-0.9875*10^{7}$
24.	$0.2*10^{6}$	$0.23*10^{6}$	$0.235*10^{6}$	$0.2355*10^{6}$
25.	$0.2*10^{-11}$	$0.24*10^{-11}$	$0.249*10^{-11}$	$0.2493*10^{-11}$

2.3
1.

Decimal	Binary	Octal	**Hexadecimal**
0	0	0	0
1	1	1	1
2	10	2	2
3	11	3	3
4	100	4	4
5	101	5	5
6	110	6	6
7	111	7	7
8	1000	10	8
9	1001	11	9
10	1010	12	A
11	1011	13	B

Decimal	Binary	Octal	Hexadecimal
12	1100	14	C
13	1101	15	D
14	1110	16	E
15	1111	17	F
16	10000	20	10
17	10001	21	11
18	10010	22	12
19	10011	23	13
20	10100	24	14
21	10101	25	15
22	10110	26	16
23	10111	27	17
24	11000	30	18
25	11001	31	19
26	11010	32	1A
27	11011	33	1B
28	11100	34	1C
29	11101	35	1D
30	11110	36	1E
31	11111	37	1F
32	10 0000	40	20
33	10 0001	41	21
34	10 0010	42	22
35	10 0011	43	23
36	10 01001	44	24

2. 2^3
3. 2^5
4. a. 32 or 2^5 b. 64 or 2^6 c. 128 or 2^7 d. 512 or 2^9
5. a. 32 or 2^5 b. 128 or 2^7 c. 256 or 2^8 d. 128 or 2^7
6. 12
7. 47
8. 36
9. 3730
10. 534
11. 3072
12. 39,026
13. 43,981
14. 14,533
15. 28,236
16. 167
17. 279
18. 878
19. 142
20. 2322

	base 2	base 8	base 16
21.	10 0101	45	25
22.	1110 1111	357	EF

	base 2	base 8	base 16
23.	1 0000	20	10
24.	1110 0011	343	E3
25.	11 0111	67	37
26.	1 0010	22	12

2.4

1.

4	2	9	.	3	1
10^2	10^1	10^0		10^{-1}	10^{-2}
100	10	1		.1	.01
400	20	9		.3	.01

2.

5	8	8	.	1	3	5	6
10^2	10^1	10^0		10^{-1}	10^{-2}	10^{-3}	10^{-4}
100	10	1		.1	.01	.001	.0001
500	80	8		.1	.03	.005	.0006

3.

2	3	9	7	.	1	5
10^3	10^2	10^1	10^0		10^{-1}	10^{-2}
1000	100	10	1		.1	.01
2000	300	90	7		.1	.05

4.

8	9	9	.	9	0	1
10^2	10^1	10^0		10^{-1}	10^{-2}	10^{-3}
100	10	1		.1	.01	.001
800	90	9		.9	.00	.001

5. 9.1875
6. 74.5
7. 210.03125
8. 31.9375
9. 5.625
10. 29.375
11. 15.9375
12. 3.125
13. 430.06640625
14. 238.9296875
15. 2748.791748046
16. 77
17. 212.5
18. 385.25
19. 52.5625
20. 23.875

2.5

1. 100100011_2
2. 5544_8
3. $9DC_{16}$
4. 100111100_2
5. 10550_8
6. $B90_{16}$
7. 1000111_2
8. 1000100_2
9. 4042_8
10. $1B2_{16}$
11. 2767_{16}
12. 733, 731

13. 1235, 1233
14. 1110_2, 1100_2
15. 10100111_2, 101001010_2
16. 234, 236
17. 1000_8, 776_8
18. $92B_{16}$, 929_{16}
19. $F30_{16}$, $F2E_{16}$

2.6
1. $0101\ 1011_2$
2. $1010\ 0101_2$
3. $0111\ 0011_2$
4. $1101\ 1011_2$
5. Overflow
6. Overflow
7. $0100\ 0001_2$
8. $0000\ 0000\ 0101\ 1011_2$
9. $1111\ 1111\ 1010\ 0101_2$
10. $1111\ 1111\ 0110\ 1001_2$
11. $1000\ 1000\ 1000\ 1001_2$
12. Overflow
13. $1010\ 0111\ 1110\ 0001_2$
14. -2,147,483,648 to 2,147,483,647
15. $-9.22*10^{18}$ to $9.22*10^{18}-1$
16. $11\ 0011_2$ which is -13
17. $11\ 1011_2$ which is -5
18. $10\ 1101_2$ which is -19
19. $10\ 1011_2$ which is -21
20. Not possible since -45 cannot be represented using 6 bits
21. 1110_2 which is -2
22. Not possible since the result should be 14 but 14 cannot be represented in a 4-bit pattern
23. 1100_2 which is -4
24. 0001_2 which is 1
25. $00\ 0010_2$ which is 2
26. $00\ 0110_2$ which is 6
27. Not possible since the result is +40 but it cannot be represented using 6 bits
28. $01\ 1000_2$ which is 24

2.7
1. ^Q
2. C
3. ^
4. Y
5. m
6. ^O
7. ;
8. L
9. p
10. G
11. ^H
12. none
13. ?
14. none
15. none
16. ^@

Chapter 3
3.1
1. statement, false
2. statement, true
3. not a statement
4. statement, true
5. statement, false
6. statement, true
7. statement, true
8. not a statement
9. not a statement
10. not a statement
11. 5 = 7
12. "Mom" < "Mommy"
13. simple statement: An integer is odd. next simple statement: The square of that integer is also an odd integer
14. no simple statements

15. simple 16. compound 17. compound 18. simple 19. simple
20. I will get a 90 on this exam nd I will get an A in the course
21. If I will get a 90 on this exam then I will take another math course
22. If I will get a 90 on this exam or I will get an A in the course, then I will take another math course.
23. If I will take another math course then I will get a 90 on this exam
24. p=I drive to school
    q=I will need to have a driver's license
    if p, then q
25. p=You are drinking and driving
    q=You are breaking the law
    if p, then q
26. p=You semester g.p.a. is at least 3.5
    q=You carry at least a 12 credit load
    r=You are eligible for the Dean's List
    if p and q, then r
27. simple statements: $10 > 15$
    $7 < 20$
    connective: AND
28. simple statements: $(5 + 3) > 7$
    $(12 + 4) < 10$
    connective: OR
29. simple statementS: $-5 < 9$
    $2 = 11$
    connective: AND
30. simple statements: $7 < 15$
    $16 = 32$
    connectives: OR, AND

3.2
1. p=The computer is on
   q=The printer is broken
   p ∨ q
2. p=Today is a holiday
   q=Classes are cancelled
   p − q
3. p=Mary is the class president
   q=She likes classical music
   p −q
4. p=Sue and Dan are coming to dinner
   q=We are going to watch television
   p ∨ q
5. p=You study for the test
   q=You will fail the course
   p ∨ q
6. p=There is a storm
   q=The electricity went out
   p − q
7. false
8. false
9. p − r evalutes to false, so (p − r) ∨ q evaluates to false
10. Information Technology students take MA17 and MA17 is a corequisite of Visual Basic programming
11. Information Technology students take MA17 or MA17 is a corequisite of Visual Basic programming
12. p is false, q is false, and r is true, so (p − r) ∨ q is false

13. p is false, q is false, and r is true, so $(q - r) \vee p$ is true

3.3

1. $\sim p$: No computers are outdated. false
2. $\sim q$: $-5 < 12$ true
3. $\sim r$: $24 \leq 15$ false
4. $\sim s$: Not all students in this class are Information Technology majors true
5. $\sim t$: Every integer is the solution to the equation $x^2 > -8$ false
6. $\sim w$: Selden is not the capital of New York state true
7. $\sim x$: The Yankees are a foorball team. false
8. $\sim y$: $6 \geq 10$ false
9. $\sim z$: Not all natural numbers are positive false
10. $\sim m$: All real numbers are rational numbers false
11. Mary takes discrete mathematics and Mary is a computer science major
12. Mary takes discrete mathematics or Mary is a computer science major
13. Mary does not take discrete mathematics and Mary is not a computer science major
14. Mary takes discrete mathematics or Mary is not a computer science major
15. It is not true that Mary takes discrete mathematics and is a computer science major
16. $(3 \leq 5) - (8 > 0)$ true
17. $(3 \leq 5) \vee (8 > 0)$ true
18. $(3 \leq 5) \vee (8 \leq 0)$ true
19. From number sixteen we know that $p - q$ is true, so $\sim (p - q)$ must be false
20. From number seventeen, we know that $p \vee q$ is true, so $\sim (p \vee q)$ must be false

3.4

1.

p	q	~q	$p \wedge \sim q$
T	T	F	F
T	F	T	T
F	T	F	F
F	F	T	F

2.

p	q	~p	~p ∨ q	(~p ∨ q) ∨ p
T	T	F	T	T
T	F	F	F	T
F	T	T	T	T
F	F	T	T	T

3.

p	q	~p	p − q	~p ∨	(p ∧ q) ∨ (~p ∨
T	T	F	T	T	T
T	F	F	F	F	F
F	T	T	F	T	T
F	F	T	F	T	T

4.

p	q	r	~q	~q − r	p ∨ (~q ∧ r)
T	T	T	F	F	T
T	T	F	F	F	T
T	F	T	T	T	T
T	F	F	T	F	T
F	T	T	F	F	F
F	T	F	F	F	F
F	F	T	T	T	T
F	F	F	T	F	F

5.

p	q	p ∨ q	~(p ∨ q)
T	T	T	F
T	F	T	F
F	T	T	F
F	F	F	T

6.

p	q	r	~p	~r	q→~r	~p ∨ (q ∧ ~r)
T	T	T	F	F	F	F
T	T	F	F	T	T	T
T	F	T	F	F	F	F
T	F	F	F	T	F	F
F	T	T	T	F	F	T
F	T	F	T	T	T	T
F	F	T	T	F	F	T
F	F	F	T	T	F	T

7.

p	q	r	s	~r	p→q	~r→s	(p→q) ∨ (~r ∧ s)
T	T	T	T	F	T	F	T
T	T	T	F	F	T	F	T
T	T	F	T	T	T	T	T
T	T	F	F	T	T	F	T
T	F	T	T	F	F	F	F
T	F	T	F	F	F	F	F
T	F	F	T	T	F	T	T
T	F	F	F	T	F	F	F
F	T	T	T	F	F	F	F
F	T	T	F	F	F	F	F
F	T	F	T	T	F	T	T
F	T	F	F	T	F	F	F
F	F	T	T	F	F	F	F
F	F	T	F	F	F	F	F
F	F	F	T	T	F	T	T
F	F	F	F	T	F	F	F

8. True 9. True 10. False 11. True 12. False

For exercises 13-16, we use the following truth values: p=T, q=F, r=T

13. False 14. True 15. True 16. True
17. Peter is an information technology major or Peter is taking MA17.
18. Peter is not an information technology major.
19. Peter is an information technology major and Peter is not taking MA17.
20. Peter is not an information technology major and Peter is taking MA17.
21. Peter is not an information technology major or Peter is not taking MA17.

3.5

1.

p	q	$\sim p$	$\sim q$	$\sim p \Rightarrow \sim q$
T	T	F	F	T
T	F	F	T	T
F	T	T	F	F
F	F	T	T	T

2.

p	q	$\sim q$	$p \Rightarrow q$	$(p \Rightarrow q) \wedge \sim q$
T	T	F	T	F
T	F	T	F	F
F	T	F	T	F
F	F	T	T	T

3.

p	q	$\sim q$	$\sim q \vee p$	$(\sim q \vee p) \Rightarrow p$
T	T	F	T	T
T	F	T	T	T
F	T	F	F	T
F	F	T	T	F

4.

p	q	r	$p \Rightarrow q$	$(p \Rightarrow q) \Rightarrow r$
T	T	T	T	T
T	T	F	T	F
T	F	T	F	T
T	F	F	F	T
F	T	T	T	T
F	T	F	T	F
F	F	T	T	T
F	F	F	T	F

5.

p	q	$\sim q$	$p \Leftrightarrow \sim q$	$p \vee q$	$(p \Leftrightarrow \sim q) \Rightarrow (p \vee q)$
T	T	F	F	T	T
T	F	T	T	T	T
F	T	F	T	T	T
F	F	T	F	F	T

6.

p	q	r	$q \Rightarrow r$	$p \Rightarrow (q \Rightarrow r)$
T	T	T	T	T
T	T	F	F	F
T	F	T	T	T
T	F	F	T	T
F	T	T	T	T
F	T	F	F	T
F	F	T	T	T
F	F	F	T	T

7. true
8. true
9. true
10. false
11. true
12. false

For exercises 13-16, we use the following truth values: p=T, q=F, r=T

13. false
14. true
15. false
16. false

17. (a) more than 9
    (b) Less than or equal to 9

18. (a) no output
    (b) Hi
    (c) Hi

19. (a) You can vote in the upcoming election
    (b) You can vote in 2 years

20. (a) your grade is an F
    (b) Your grade is a C
    (c) Your grade is a B

3.6

1.

p	q	$\sim p$	$\sim q$	$p \vee q$	$\sim (p \vee q)$	$\sim p \wedge \sim q$
T	T	F	F	T	F	F
T	F	F	T	T	F	F
F	T	T	F	T	F	F
F	F	T	T	F	T	T

2.

p	q	r	$q \vee r$	$p \wedge (q \vee r)$	$p \wedge q$	$p \wedge r$	$(p \wedge q) \vee (p \wedge r)$
T	T	T	T	T	T	T	T
T	T	F	T	T	T	F	T
T	F	T	T	T	F	T	T
T	F	F	F	F	F	F	F
F	T	T	T	F	F	F	F
F	T	F	T	F	F	F	F
F	F	T	T	F	F	F	F
F	F	F	F	F	F	F	F

3.

p	q	~p	p ⇒ q	~p ∧ q
T	T	F	T	F
T	F	F	F	F
F	T	T	T	T
F	F	T	T	F

4.

p	q	r	q ∨ r	p ⇒ (q ∨ r)	p ⇒ q	p ⇒ r	(p ⇒ q) ∨ (p ⇒ r)
T	T	T	T	T	T	T	T
T	T	F	T	T	T	F	T
T	F	T	T	T	F	T	T
T	F	F	F	F	F	F	F
F	T	T	T	T	T	T	T
F	T	F	T	T	T	T	T
F	F	T	T	T	T	T	T
F	F	F	F	T	T	T	T

5.

p	q	~q	p ⇒ q	~(p ⇒ q)	p ∧ ~q
T	T	F	T	F	F
T	F	T	F	T	T
F	T	F	T	F	F
F	F	T	T	F	F

We can see that exercises 1, 2, 4, and 5 have statements which are logically equivalent.

6. It is false that roses are red and it is false that the leaves are green

7. $3 \leq 5$ OR $8 \geq 12$

8. $x \neq 4$ AND $y \geq 10$

9. (a) contrapositive: If you did not violate the traffic rules, then you don't get a ticket.
   (b) converse: If you violated the traffic rules, then you get a ticket.
   (c) inverse: If you don't get a ticket, then you don't violate the traffic rules

10. (a) contrapositive: If it is not raining, then today is not Saturday
    (b) converse: If it is raining, then today is Saturday
    (c) inverse: If today is not Saturday, then it is not raining

11. (a) contrapositive: If it does not fly, then it has no wings
    (b) converse: If it flies, then it has wings
    (c) inverse: If it has no wings, then it does not fly

12. (a) contrapositive: If you are not on the Dean's List, then it is false that your GPA is 3.5 or higher and you carry a twelve-credit load
    (b) converse: If you are on the Dean's List, then your GPA is 3.5 or higher and you carry a twelve-credit load
    (c) inverse: If it is false that your GPA is 3.5 or higher and you carry a twelve-credit load, then you are not on the Dean's List

13. If $2 < 4$, then $3 > 5$. false

    (a) contrapositive: If $3 \leq 5$, then $2 \geq 4$. false
    (b) converse: If $3 > 5$, then $2 < 4$. true
    (c) inverse: If $2 \geq 4$, then $3 \leq 5$. true

14. If $6 > 8$, then $9 < 11$. true

    (a) contrapositive: If $9 \geq 11$, then $6 \leq 8$. true
    (b) converse: If $9 < 11$, then $6 > 8$. false
    (c) inverse: If $6 \leq 8$, then $9 \geq 11$. false

15.

p	$\sim p$	$p \vee \sim p$
T	F	T
T	F	T
F	T	T
F	T	T

3.7

1.

p	q	r	$p \wedge q$	$(p \wedge q) \vee r$
1	1	1	1	1
1	1	0	1	1
1	0	1	0	1
1	0	0	0	0
0	1	1	0	1
0	1	0	0	0
0	0	1	0	1
0	0	0	0	0

2.

p	q	r	$\sim p$	$r \vee q$	$\sim p \wedge (r \vee q)$
1	1	1	0	1	0
1	1	0	0	1	0
1	0	1	0	1	0
1	0	0	0	0	0
0	1	1	1	1	1
0	1	0	1	1	1
0	0	1	1	1	1
0	0	0	1	0	0

3.

p	q	r	$\sim r$	$p \wedge q$	$(p \wedge q) \vee \sim r$	$p \vee r$	$[(p \wedge q) \vee \sim r] \wedge (p \vee r)$
1	1	1	0	1	1	1	1
1	1	0	1	1	1	1	1
1	0	1	0	0	0	1	0
1	0	0	1	0	1	1	1
0	1	1	0	0	0	1	0
0	1	0	1	0	1	0	0
0	0	1	0	0	0	1	0
0	0	0	1	0	1	0	0

4.

p	q	$\sim p$	$\sim p \vee q$
1	1	0	1
1	0	0	0
0	1	1	1
0	0	1	1

5.

p	q	r	$\sim p$	$p \wedge q$	$r \wedge \sim p$	$(p \wedge q) \vee (r \wedge \sim p)$
1	1	1	0	1	0	1
1	1	0	0	1	0	1
1	0	1	0	0	0	0
1	0	0	0	0	0	0
0	1	1	1	0	1	1
0	1	0	1	0	0	0
0	0	1	1	0	1	1
0	0	0	1	0	0	0

6.

p	q	r	$\sim p$	$r \vee q$	$r \vee \sim p$	$(r \vee q) \wedge (r \vee \sim p)$
1	1	1	0	1	1	1
1	1	0	0	1	0	0
1	0	1	0	1	1	1
1	0	0	0	0	0	0
0	1	1	1	1	1	1
0	1	0	1	1	1	1
0	0	1	1	1	1	1
0	0	0	1	0	1	0

7.

p	q	r	$\sim p$	$\sim p \wedge q$	$q \wedge r$	$(\sim p \wedge q) \vee (q \wedge r)$
1	1	1	0	0	1	1
1	1	0	0	0	0	0
1	0	1	0	0	0	0
1	0	0	0	0	0	0
0	1	1	1	1	1	1
0	1	0	1	1	0	1
0	0	1	1	0	0	0
0	0	0	1	0	0	0

8.

p	q	r	$\sim q$	$p \vee \sim q$	$q \wedge r$	$p \Rightarrow q$	$(p \vee \sim q) \vee (q \wedge r)$	$[(p \vee \sim q) \vee (q \wedge r)] \vee (p \Rightarrow q)$
1	1	1	0	1	1	1	1	1
1	1	0	0	1	0	1	1	1
1	0	1	1	1	0	0	1	1
1	0	0	1	1	0	0	1	1
0	1	1	0	0	1	1	1	1
0	1	0	0	0	0	1	0	1
0	0	1	1	1	0	1	1	1
0	0	0	1	1	0	1	1	1

9.

p	q	r	~p	~q	~p⇒q	r∧~q	(~p⇒q)⇒(r∧~q)
1	1	1	0	0	1	0	0
1	1	0	0	0	1	0	0
1	0	1	0	1	1	1	1
1	0	0	0	1	1	0	0
0	1	1	1	0	1	0	0
0	1	0	1	0	1	0	0
0	0	1	1	1	0	1	1
0	0	0	1	1	0	0	1

10.

p	q	p⇒q	q⇒p	(p⇒q)∧(q⇒p)
1	1	1	1	1
1	0	0	1	0
0	1	1	0	0
0	0	1	1	1

Chapter 4
4.1

1. A relation is a rule that takes an input and produces one or more outputs.

2. A function is a rule that takes an input and produces only one output.

3. All functions are relations since it takes an input and produces an output.

4. All relations are not functions because the input can produce one or more outputs, while a function only produces one and not more than one.

5. The domain of a function is the input while the range is the output.

6. Consider the relation between a person and their bank accounts. This relation is not a function since a person can have one or more bank accounts. For example, Jill can have a Fleet bank account while Jack has a bank account at Fleet, Washington Mutual, and Citibank.

7. Refer to diagrams given out in class

8. Refer to diagrams given out in class

9. Part a is not a function because there are inputs which map to multiple outputs. For example, Tom Hanks maps to both Sleepless in Seattle and Saving Private Ryan.
Part b and c are functions because each input produces only one output. In part c, it is ok that Sue and Danise map to the same number of siblings (it is still considered a function).

10. (a) $f(2) = -2(2) + 5 = -4 + 5 = 1$
    (b) $f(-3) = -2(-3) + 5 = 6 + 5 = 11$

11. (a) $g(0) = -2(0)^2 + 5(0) - 3 = -3$
    (b) $g(-1) = -2(-1)^2 + 5(-1) - 3 = -2 - 5 - 3 = -10$

12. (a) $f(0) = \frac{0^2 - 1}{0 - 2} = \frac{1}{2}$
    (b) f(2) is undefined since zero is in the denominator
    (c) $f(-2) = \frac{(-2)^2 - 1}{-2 - 2} = \frac{4 - 1}{-4} = -\frac{3}{4}$

13. (a) $g(3) = \frac{1}{3}(3)^2 - 3 = \frac{9}{3} - 3 = 3 - 3 = 0$
    (b) $g(1) = \frac{1}{3}(1)^2 - 1 = \frac{1}{3} - 1 = -\frac{2}{3}$

14. (a) $C(F) = \frac{5}{9}(F - 32)$
    (b) $C(0) = \frac{5}{9}(0 - 32) = \frac{5}{9}(-32) = -\frac{160}{9} \approx -17.78$
    $C(-10) = \frac{5}{9}(-10 - 32) = \frac{5}{9}(-42) = \frac{-210}{9} \approx 23.33$
    $C(82) = \frac{5}{9}(82 - 32) = \frac{5}{9}(50) = \frac{250}{9} \approx 27.78$
    $C(212) = \frac{5}{9}(212 - 32) = \frac{5}{9}(180) = 100$
    $C(32) = \frac{5}{9}(32 - 32) = 0$
    $C(29) = \frac{5}{9}(29 - 32) = \frac{5}{9}(-3) = \frac{-15}{9} \approx -1.67$

15. $I(.045) = 10,000(.045) = 450$
    $I(.0725) = 10,000(.0725) = 725$
    $I(.197) = 10,000(.197) = 1970$

16. $I(.0399) = 500(.0399) = 19.95$
    $I(.0799) = 500(.0799) = 39.95$
    $I(.1899) = 500(.1899) = 94.95$
    $I(.2399) = 500(.2399) = 119.95$

17. (a) $H(45) = 100 - \frac{45}{5} = 100 - 9 = 91$
    $H(450) = 100 - \frac{450}{5} = 100 - 90 = 10$
    $H(212) = 100 - \frac{212}{5} = 100 - 42.4 = 57.6$
    $H(625) = 100 - \frac{625}{5} = 100 - 125 = -25$

(b) Let us first determine the number of shots required before a player's health level changes from Excellent to Good. We know that a player's health is rated Excellent if H is greater than or equal to 85. We can determine the number of times the player got hit by solving for x. That is,
$H(x) = 85 = 100 - \frac{x}{5}$
$-15 = -\frac{x}{5}$
$x = 75$

This means that the most number of times a player could get hit and still have a health level of Excellent is 75 shots. That is, for the health level to change from Excellent to Good, the number of shots must be 76 or more. We follow the same reasoning for the other level changes.

Good to Fair is 176 shots, Fair to Poor is 301 shots, and Poor to Spirit is 500 shots.

4.2

1. 5
2. 3
3. 4
4. undefined
5. 3
6. 4
7. 0
8. 3
9. 1
10. 4.8987 …
11. $\log_2(8) = 3$
12. $\log_2(64) = 6$
13. $\log_3(9) = 2$
14. $\log_2(1/8) = -3$
15. $\log_5(1/625) = -4$
16. $\log_{81}(9) = ½$
17. $2^3 = 8$
18. $5^3 = 125$
19. $81^{½} = 9$
20. $e^3 = e^3$
21. $10^{-2} = 1/100$
22. $2^{-6} = 1/64$
23. $3^3 = x$
24. 5
25. 3
26. 3
27. -2
28. -4
29. -1
30. -3
31. 1.7918
32. 1.6335
33. 3.0103
34. 15.5885
35. 0.0498
36. 0.000976 or 0.0010
37. $\log_3(4) + \log_3(10)$
38. $\log_2(7) + \log_2(x+2)$
39. $\ln(5) - \ln(19)$
40. $\log_8(x) - \log_8(x+1)$
41. $7\log_2(2)$ or 7
42. $-3\log_5(5)$ or -3
43. $3\log_4(3) + 2\log_4(x-1)$
44. $½ \log_2(x^2+1)$
45. $\log_3(8)$
46. $\log_2(5x)$
47. $\log_{10}(5)$
48. $\log_3(4)$
49. $\log_2(3^3 x^2)$
50. $\log_4(81)^{½}$ or $\log_4(9)$

4.6
3.
a. 24
b. 5040
c. 479001600
d. 89
e. undefined
f. 377

Chapter 5
5.1
For exercises 1-5, we determine the size of the vector by counting the number of elements in the vector.

1. 4
2. 6
3. 4
4. 7
5. 3

The pair of vectors are equal if and only if their sizes are equal and their corresponding elements are equal

6. yes
7. no
8. no
9. yes
10. no

5.2
1. [5 6]
2. [2 10 -1]
3. [-1 -1]
4. [4 -9 10]
5. [13 -7]

6. $a\mathbf{u}$=[8 -4 6]
7. $b\mathbf{v}$=[-3 21 15]
8. $a\mathbf{u}+b\mathbf{v}$=[5 17 21]
9. $\mathbf{v}-\mathbf{u}$=[-5 9 2]
10. $b\mathbf{v}-\mathbf{u}$=[-7 19 12]
11. $\mathbf{v}+\mathbf{w}$=[-4 17 -2]
12. $\mathbf{u}\cdot\mathbf{v}$=-3
13. $(2\mathbf{w})\cdot\mathbf{v}$=76
14. $\mathbf{u}\cdot\mathbf{w}$=-53
15. $a(\mathbf{u}+\mathbf{w})$=[2 16 -8]
16. –14

17. Cannot multiply two vectors because number of row vector columns does not equal the number of column vector rows
18. 55
19. Cannot multiply two vectors – see exercise two
20. 4

5.3
1. 2x3
2. 3x3
3. 1x4
4. 5x2
5. P_{22}=6
6. P_{43}=3
7. P(4,1)=4
8. P(1,3)=0
9. x=7, y=4
10. w=1, x=2, y=3, z=-3
11. The matrices cannot be equal because they are not of the same order
12. x=2, y=3, z=5
13. x=4, y=3, z=5

5.4

1. $2\mathbf{A} = \begin{bmatrix} 4 & -6 & 8 \\ -2 & 4 & -10 \end{bmatrix}$

2. $-3\mathbf{C} = \begin{bmatrix} -3 & 0 & 3 \\ 9 & -6 & -12 \\ 0 & -21 & 0 \end{bmatrix}$

3. $\mathbf{A}+\mathbf{B} = \begin{bmatrix} 3 & -3 & 1 \\ -3 & 3 & -2 \end{bmatrix}$

4. $\mathbf{A}-2\mathbf{B} = \begin{bmatrix} 4 & -3 & -2 \\ -5 & 4 & 1 \end{bmatrix}$

5. **A+C** – cannot add because the matrices are not of the same order

6. $-\mathbf{D} = \begin{bmatrix} -6 & 3 & -1 \\ 2 & 0 & 3 \\ 0 & 1 & -7 \end{bmatrix}$

7. $2\mathbf{C}+\mathbf{D} = \begin{bmatrix} 8 & -3 & -1 \\ -8 & 4 & 5 \\ 0 & 13 & 7 \end{bmatrix}$

8. **D-B** – cannot add because the matrices are not of the same order
9. **AB** – **A**'s number of columns is not equal to **B**'s number of rows
10. $\mathbf{BD} = \begin{bmatrix} 6 & 0 & -20 \\ -14 & 3 & 16 \end{bmatrix}$
11. **CA** – **C**'s number of columns is not equal to **A**'s number of rows
12. $\mathbf{CD} = \begin{bmatrix} 6 & -2 & -6 \\ -22 & 5 & 19 \\ -14 & 0 & -21 \end{bmatrix}$
13. **BA** – **B**'s number of columns is not equal to **A**'s number of rows
14. **(2A)·B** – **A**'s number of columns is not equal to **B**'s number of rows
15. **DB** – **D**'s number of columns is not equal to **B**'s number of rows
16. $\mathbf{CD} = \begin{bmatrix} 6 & -2 & -6 \\ -22 & 5 & 19 \\ -14 & 0 & -21 \end{bmatrix}$
17. $\mathbf{DC} = \begin{bmatrix} 15 & 1 & -18 \\ -2 & -21 & 2 \\ 3 & 47 & -4 \end{bmatrix}$
18. Matrix multiplication is not commutative because the results obtained are not the same.

5.5

1. 24503500
2. 846
3. 116403
4. 24183900
5. 935
6. 290
7. 34
8. 1495
9. 214
10. 684
11. 110
12. 315
13. 720
14. 16384
15. 7560
16. $72/5 = 14.4$
17. $1/3 = .3333$
18. 576
19. $1/5 = .2$
20. 21

Chapter 6
6.1

1. S={1,2,3,4,5,6,7,8,9}
2. {-3,-2,-1,0,1,2,3,4,5}
3. A={-2,0,2,4,6,8,10,12,14,16}
4. {16,18,20,22,24,26,28}
5. {7,9,11,13,15,17,19,21}
6. {12}
7. {}
8. {-5,5}
9. {1,2,3,4,5,6}
10. {-2,-1,0,1,2,3,4,5,6,7,8,9,10}
11. {x | x is a number divisible by 5 and $0 \le x \le 20$}
12. { x | x is an even number between 2 and 30 inclusive}
13. { x | x ∈ N and is an even number}
14. { x | x ∈ N and is an odd number }
15. { x | x ∈ Z and is an even number }
16. ∈
17. ∉
18. ∈
19. ∉
20. ∈
21. ∉
22. ∉
23. ∉
24. ∉
25. ∈
26. Well-defined, it may be difficult but possible to count the number of people
27. Not well defined, must define a what height qualifies as a big buildings
28. Not well defined, each student has a different view on if a course is fun or not
29. well defined, there are only a finite number of numbers that are in the hexadecimal number system
30. Not well defined, does not define what a fast processor is

6.2
1. ⊆ ⊂
2. ⊄
3. ⊆ ⊂
4. ⊆ ⊂
5. ⊆ ⊂
6. ⊆ ⊂
7. ⊂
8. ⊆ ⊂
9. ⊂
10. ⊆ ⊂
11. Ā={a, b, c d}
12. $\overline{C}$={d, e, f, g}
13. $\overline{\overline{S}}$={a, c, e}
14. $\overline{\overline{B}}$={a, b, c, d, e, f, g}
15. 4 subsets: {}, {cat}, {dog}, {cat, dog}
16. 8 subsets: {},{1},{2},{3},{1,2},{2,3},{1,3},{1,2,3}
17. 2 subsets: {}, {19}
18. 4 subsets: {}, {0}, {1}, {0, 1}
19. 16 subsets: {}, {a}, {b}, {c}, {d}, {a, b}, {b, c}, {c, d}, {a, d}, {a, c}, {b, d}, {a, b, c}, {b, c, d}, {a, c, d}, {a, b, d}, {a, b, c, d}
20. 1 subset: {}
21. 3 subsets: {}, {cat}, {dog
22. 7 subsets: {},{1},{2},{3},{1,2},{2,3},{1,3}
23. 1 subset: {}
24. 3 subsets: {}, {0}, {1}
25. 15 subsets: {}, {a}, {b}, {c}, {d}, {a, b}, {b, c}, {c, d}, {a, d}, {a, c}, {b, d}, {a, b, c}, {b, c, d}, {a, c, d}, {a, b, d}
26. 0 subsets

27.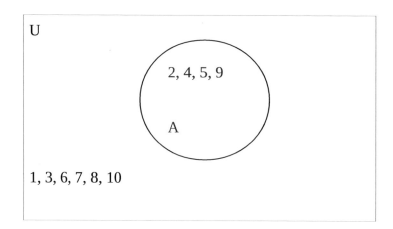

28.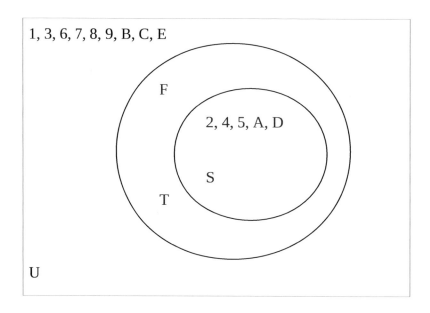

29. a.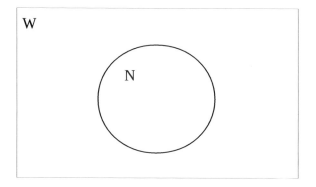

b.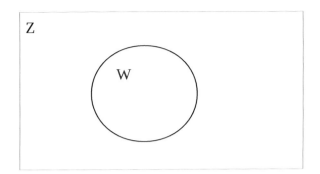

c.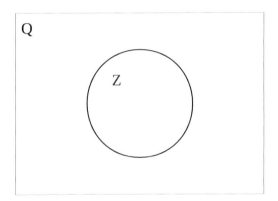

d.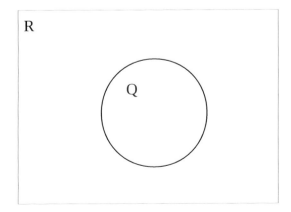

e.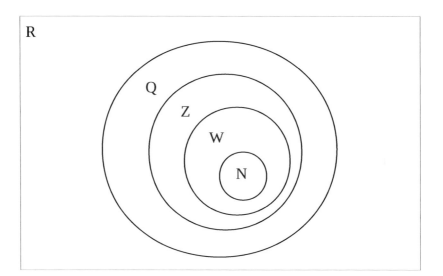

Made in the USA
Middletown, DE
03 February 2019